Virtual Data Warehousing, Data Publishing, and Call Detail

David Belanger, Kenneth Church, and Andrew Hume

AT&T InfoLab
AT&T Labs Research
180 Park Ave.
Florham Park, NJ, USA
{dgb, kqc, andrew}@research.att.com

Abstract. Many telecommunication applications can be modeled as data flowing through a network of systems. For example, the billing factory can be thought of as records of telephone calls flowing from one process to the next, as in from recording to rating to rendering. Probes can be attached at various points in the network. On the basis of these probes we would like to reason about the flow as a whole. Are there any losses? If so, where? How long does it take for a record to flow from one step to the next? Where are the bottlenecks? Two systems are described in a dataflow context: (1) Gecko, a system monitoring a very large billing system and (2) Pzip: a general compression tool based on gzip, which is typically twice as good as gzip in both space and time for sequences of fixed-length records.

1 Introduction

Many telecommunication applications can be modeled as data flowing through a network of systems. For example, the AT&T billing factory can be thought of as records of telephone calls flowing from one process to the next, e.g., from recording to rating to rendering. We would like to reason about (query) the flow as a whole. How many packets are being lost? If so, where? How long does it take for a packet to flow from here to there? Where are the major bottlenecks? We introduce the term *dataflow* to refer to an abstract process of data flowing through a network. Questions about throughput, latency, loss, error rates and so forth can be addressed by attaching probes (packet sniffers) at various points in the network. An alerting system is a simple example. There is a single probe that counts events, such as error conditions, and computes a summary statistic such as a running estimate of the average error rate. This figure of merit is then compared to a threshold or a control limit to decide whether or not to sound an alarm. More interesting examples involve a *join* of the evidence from two or more probes. To study the latencies in billing, Hume and his colleagues (Hume and Maclellan, 1999) built a system called *Gecko* that taps the flow in the billing systems at roughly a dozen locations and co⎽ see how long it takes for records to propagate fro

W. Jonker (Ed.): Databases in Telecommunications, LNCS 18⎽

What makes dataflows challenging is the combination of scale and timeliness. The networks we want to monitor such as the billing systems tend to be quite significant in a number of dimensions: throughput, latency, reliability, physical distance and cost. But if we are going to tap this network at multiple places and make inferences across these tap points, it is likely that the measurement network will require even more capacity in terms of bandwidth and/or state than the underlying network that is being monitored. We define a dataflow as an abstract database of triples, $\langle p, t, l \rangle$, that keeps track of all packets, p, over all times, t, and all locations, l. We would like to query this database to make inferences over time and location. Obviously, we can't materialize the database. The probes provide direct evidence for a few of these $\langle p, t, l \rangle$ triples (e.g., the packets that have been received thus far), but many of them (especially the packets that will be received in the future) will have to be inferred by indirect statistical methods. We view the evidence from the probes as a training sample, based on which we hope to derive inferences that will generalize to unseen data. The distinction between training data and test data will be especially important in the discussion of compression below. Dataflows borrow many ideas from conventional databases, especially data warehouses and transaction databases. But dataflows emphasize the movement of the data in time and/or space. Suppose that we wanted to *join* (or merge/purge as in Hernandez, M. and Stolfo, 1995) the data being emitted by two I/O streams? Would it be better to use an indexed database or Unix pipes? Indexing can be very helpful, especially when the data are relatively static (fewer loads than queries). But indexing can be a liability when the data are flowing quickly. If there are more loads than queries, then the database could easily end up spending the bulk of its computational resources creating indexes that will never be used. In the two applications that will be described below, we ended up with a load and *join* process that is somewhere between a database and a buffered I/O stream. Over the last four years, we have worked on a large number of dataflow applications in quite a number of areas including fraud, marketing, billing and customer care. The fraud applications (Cortes and Pregibon, 1998, 1999) are the most similar to alerting systems. The fraud systems look for suspicious purchasing behavior by comparing a customer's recent purchases with a summary of their long-term history (signature). As mentioned above, dataflow applications are challenging because of the scale and timeliness requirements. The fraud systems process a flow of 10 billion calls per month as quickly as possible, hopefully while there is still time to do something about it. We have also built tracking systems for marketing, which compare sales across time and space, among many other variables. Marketing is constantly trialing new ideas: a new promotion here, a new price-point there, and so on. They will do more of what works and less of what doesn't. The challenge is to find ways to distinguish what works from what doesn't as quickly as possible, hopefully while there is still time to respond. This paper will discuss two dataflow applications in more detail. The first example emphasizes scale (high throughput with low latency) while the second example makes predictions from training data to unseen test data.

- *Gecko*: a system that monitors the billing factory (Hume and Maclellan, 1999). How long does it take for a record to flow from one system to the next? Do any records get lost along the way? Latencies in the billing factory can be thought of as inventory and lost records can be thought of as shrinkage, a term used in retailing to refer to missing/damaged inventory. As mentioned above, *Gecko* answers these questions by *joining* the views of the dataflow from a number of probes distributed throughout the billing factory.
- *Pzip*: a lossless, semantics-free compression tool based on *gzip* (`www.gzip.org`). *Gzip* has just two phases, a compression phase and a decompression phase. *Pzip* introduces a third phase, a training phase, which is given a training sample of records seen thus far and is asked to make inferences about records that will be seen in the future. In several commercially important data warehousing and data transmission applications, these inferences have improved the performance of *gzip* by at least a factor of two in both space and time.

In this paper, we use the term *dataflow architecture* to refer to a collection of probes that sample data moving quickly in time and/or space through a network. The challenge is to support as many of the features of standard databases as possible with this architecture, but hopefully without transporting all of the observations (and statistical inferences) to a central point in time and space and loading everything into a single materialized view.

2   Gecko

The *Gecko* system monitors residential calls flowing through the billing factory for residential customers (figure 1). As each call is completed, a record is generated at one of about 160 4ESS/5ESS switches distributed geographically around the USA. These calls are collected by the BILLDATS system in Arizona, and are then sent to RICS, which, after converting from the 200 or so switch formats into the appropriate biller specific format, routes them to one of 50 or so billers. We denote the 45 or so business billers collectively as BCS and are thereafter ignored. After RICS, the remaining records are sent to MPS, a large buffer that holds the records until the customers next bill cycle, approximately half a month on average (some customers are billed every second month). Every day, Monday through Saturday, the main residential biller, RAMP, requests all records for about 5The monitoring task is challenging because of the throughput and latency requirements. Each box in figure 1 has a throughput of 1-10 billion calls/month (more toward the beginning of the pipeline and less downstream of the link from RICS to BCS in figure 1). Processing of bill cycles need to be completed in a day or less. Once a day, *Gecko* updates its internal database and generates reports tracking shrinkage and inventory. On an average day, the update process handles a billion records in 8-9 hours, but on the busiest day, the process has handled 5 billion records in 13 hours.

The monitoring task requires considerably more bandwidth and state than the underlying process that is being monitored. Tapping a flow with k probes ge-

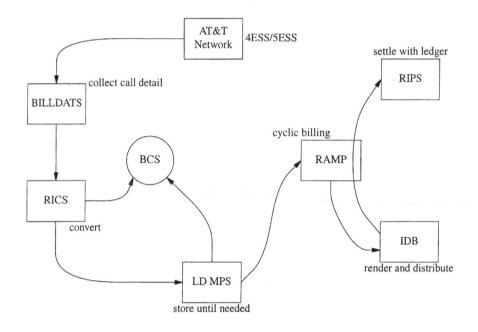

Fig. 1. The Billing Systems for Residential Customers

nerally increases the bandwidth by a factor of k. In *Gecko*, 2/3 of the records pass 3 probes and 1/3 pass about 6, for an overall average of $k = 4$. Thus, *Gecko* has about 4 times as much throughput as the residential billing factory. That is, the billion records/day mentioned above is just about 4 times the typical throughput from the switches to BILLDATS (1/4 billion records/day or 10 billion records/month).

State requirements also increase with the number of probes. *Gecko*'s database holds about 60 billion records occupying 2.6 terabytes. The underlying billing systems generally have much less state. The switches have enough state to hold onto the flow for just 2-5 days. BILLDATS and RICS have even less state, just enough for 1.5 days. *Gecko*'s database of 60 billion records is designed to store all of the flows in figure 1 for 45-75 days (the end-to-end latency in figure 1), so that they can all be reconciled with one another.

Gecko attempts to answer the question: is every call billed exactly once? As mentioned above, it does this by tapping the flow at k tap points, and *joining* the flows at each of these tap points. Although this seems an obvious approach, extreme care is required to cope with the scale and the latency requirements.

The first implementation used an off-the-shelf database, but unfortunately Hume and his colleagues found that the database could not even load the flow in real time, let alone answer the shrinkage question. In the end, they implemented the *joins* with simple sorted flat files. Loading the flow into a commercial database wasn't worth the effort since we had to release the storage as soon as possible (as soon as the *join* completed). Although a dataflow is like a database in many ways, it is also like a buffered I/O stream in the sense that we probably can't afford to store the whole flow in (secondary) memory at one time, and even if we could, we might not want to.

The *Gecko* example is interesting because it is just barely possible to bring all of the evidence to a central place. The next example of a dataflow considers a case where even this isn't practical. When compressing a large dataflow, we generally cannot afford algorithms that take superlinear time in the size of the flow, but we can afford to apply these expensive algorithms to a small sample of the flow.

3   Pin and Pzip

Pzip(<u>P</u>artition <u>Z</u>ipper) is a general purpose lossless semantics-free data compression tool for compressing sequences of fixed length records. The method is based on *gzip*, a de facto standard that has been adopted by commercial database vendors such as Sybase (Goldstein et al., 1997). *Pzip* improves on *gzip* by introducing a training phase *pin* (<u>P</u>artition <u>In</u>ducer) that induces a transform (a partition or schema) from training data that can then be usefully applied to unseen data during compression. *Pin* finds the optimal schema by calling *gzip* many times on the training sample. It would not be practical to call *gzip* so many times on the entire flow at compression time during production, but we can afford to do so on the relatively small training sample, especially when we can perform the training offline at a convenient time of our choosing.

Pin and *pzip* were developed in collaboration with a group of developers responsible for a data warehouse used by marketing. Like *Gecko*, the warehouse uses sorted flat files rather than a proper database (most queries require one linear scan over the data, so there is little advantage to indexing). When we started working with these folks, they were using *gzip*, or rather, the closely related *zlib* libraries at www.cdrom.com/pub/infozip/zlib, to compress the flat files, with surprisingly good results. The compressed files were 15 times smaller than the original files. The marketing warehouse has since converted from *gzip* to *pzip*, saving an additional a factor of two or more in both space and time over the previous *gzip* solution. A factor of two is worth tens of millions of dollars in this application alone, just counting the cost of the hardware, and ignoring the cost of staffing the computer center, which is even larger. In addition, *pzip* is being deployed in quite a number of other applications such as sending call detail (records of telephone calls) in real time over a pipe with limited bandwidth (e.g., T1). A factor of two makes it possible to do things that could not have been done otherwise.

3.1   Rules of Engagement

When we first started working on this project, it became clear that a number of solutions, which may be very attractive on technical grounds, had to be ruled out for non-technical reasons. There are a huge number of stakeholders in a large warehouse including the funding agency, development, applications programmers, production staff and many others. It would be impossible to accomplish anything if we invited all of the stakeholders to the table. But they all have veto power. The stakeholders at the table can be asked to make compromises. Everyone else should see nothing but improvements. If we do something that negatively impacts a stakeholder that is not at the table, they will veto the project.

The folks that we were working with were responsible for compression. They had been using *gzip*. There were many other parties that we were not talking with including the applications programmers (the folks who write queries) and the production staff. Many alternatives to *pzip* would have required these folks to make compromises. For example, lossy compression techniques would have required input from the applications programmers. They would need to explain to their customers why things don't always add up exactly. Similarly, semantic compression techniques (and hand-coded schemas) could cause problems in production. How would they update the schemas and compression tools when things change and the development staff have been reassigned to another project? These discussions would have taken a long time, at best, and probably would not have reached a successful conclusion. These rules of engagement severely constrain the space of feasible solutions.

3.2   Motivating Examples

Pzip improves over *gzip* by augmenting *gzip* with an automatically induced schema. The induction process (*pin*) takes a sample of training data and infers that some interactions among fields are important, but most are not. If there is a strong dependency among two fields, it makes sense to compress them together (row major order), but otherwise, it is generally better to compress them separately (column major order).

Consider the case of telephone numbers. Most telephone numbers in the United States are 10 digits long, but some international numbers are longer. The schema we were given allocates 16 bytes for all cases. Obviously, *gzip* will recover much of this space, but the induction process can do even better by splitting the 16-byte field into two fields, a 10-byte field for the typical telephone number and a 6-byte field for the exception. Compressing the two fields separately with *gzip* (or *zlib*) takes less space that it would take to compress the two fields together with the same compression tools.

In C notation, the induction process replaces the schema

```
char telephone_number1[16];
```

with the schema

```
struct telephone_number2 {char rule[10]; char exception[6];
};
```

The compression process then applies *gzip* to each of the induced fields separately.

Table 1. Factoring example

F0	F1	F2
000-000\n	000\n	000\n
...	...	...
000-999\n	000\n	999\n
001-000\n	001\n	000\n
...	...	...
999-999\n	999\n	999\n

Table 1 shows that this kind of factoring of the data can dramatically improve *gzip*, at least in terms of space. Let *F0* be a file of a million numbers from 0 to 999999, in sequential order, separated by newlines. In addition, hyphens are inserted in the middle of the numbers, as illustrated below. Let *F1* be a file of the numbers to the left of the hyphens and *F2* be a file of the numbers to the right of the hyphens.

Table 2 shows the improvement by compressing the two fields separately than jointly. If we apply *gzip* to the two fields separately, *F1+F2*, the compression ratio is 260, much better than the alternative, *F0*, where the compression ratio is merely 4. Of course, this example is contrived, but it isn't that different from telephone numbers. Partitioning in this way often improves over *gzip*, but typically by more like a factor of two than a factor of 65.

Table 2. Compression ratios

File	Raw	*Gzip*	Raw/*Gzip*
F0	8,000,000	2,126,120	4
F1	4,000,000	5,431	737
F2	4,000,000	25,299	158
F1 + F2	8,000,000	30,730	260

3.3 PIN: An Optimization Task

The encoding process applies an invertible transform to each block of data and then applies *gzip* (or *zlib*). The decoding process applies *gunzip* (or the equivalent *zlib* routines) to each block of data and then applies the inverse transform.

The partition induction tool, *pin*, looks at a sample of training data and searches a space of invertible transforms, looking for the one that will consume the minimum amount of space after compressing with *gzip*. Three classes of invertible transforms are currently considered: partitions, permutations and difference encodings. As a general rule of thumb, we want to move dependent fields near one another and move independent fields far apart. The permutations are intended to move dependent columns near one another and the partitions are intended to separate independent fields.

Both of these transforms greatly improve space, but they don't help that much with time. *Gzip* takes time that is roughly linear with the size of its input. We have found that many of the columns have very little information in them, and can be compressed reasonably well with a simple run length or difference encoding. Applying these quick-and-dirty compressors before *gzip* often speeds up *gzip* by quite a bit, and can also reduce space as well.

The partition induction tool, *pin*, searches this space of transforms or schemas, looking for the one that minimizes space. For expository convenience, let's consider just the space of partitions. The input to the training process is a sample of records (training data), and the record size, n. From this, the training process determines the number of fields, k, and the widths of each field, $W = w_1, w_2, ..., w_k$, where $\sum W = n$. The field widths W can be interpreted as a C structure of the form:

```
struct record {char field1[w1], ..., fieldk[wk];};
```

Each of these fields will then be compressed with *gzip*. Let s_1 be the size of the first field after compression with *gzip*, and s_2 be the size of the second field after compression with *gzip*, and s_k be the size of the k-th field after compression with *gzip*. Let $S = < s_1, s_2, ..., s_k >$ be a vector of these sizes. $\sum S$ can be thought of as an estimate of cross entropy. Cross entropy is a measure of how many bits it takes to store a record using a particular method (such as partitioning the data by W and then applying *gzip*). Entropy is the minimum of cross entropy over all possible methods.

The optimization task is to find the number of fields k and the vector of field widths W that minimize space, $\sum S$. This optimization can be solved with dynamic programming. Let $\langle a, b \rangle$ denote an interval of the database, starting at column a and spanning up to (but not including) column b.

Let $gzip(\langle a, b \rangle)$ denote how much space this interval takes when compressed with *gzip*. Similarly, let $pzip(\langle a, b \rangle)$ denote how much space this interval takes when compressed with *pzip*. The dynamic programming optimization splits the interval $\langle a, b \rangle$ at all possible cut points c, and finds the cut point that results in the best compression. The sequence of cut points determines the field boundaries from which it is straightforward to determine the number of fields, k, and the vector of field widths, W.

$$pzip(<a, b>) = min(gzip(\langle a, b \rangle), \underset{a < c < b}{MIN} [gzip(\langle a, c \rangle) + pzip(\langle c, b \rangle)])$$

This solution calls *gzip* $n(n-1)/2$ times, once for each interval $\langle i, j \rangle$, where $0 \le i < j < n$. It would not be practical to call *gzip* so many times on the entire flow, but we can afford to do so on a relatively small training sample.

Table 3. Marketing warehouse compression results (record sizes)

Compression	Size (byes)
No compression	752
Gzip	53
Gzip + Partitions	30
Pzip	25

Table 3 shows that this optimization reduces the space of the marketing warehouse by nearly a factor of two over *gzip*. Adding more elaborate transforms such as permutations and difference encoding transformations reduce the space somewhat further, and dramatically reduce the time, as well. Unfortunately, we don't know how to optimize those transforms in polynomial time, and therefore *pin* uses a greedy approximation.

All of the transforms are applied column-wise. Why don't we consider arbitrary ways to tile the block of data, and compress each tile one at a time? Why is there an asymmetry between the rows and the columns? Compressors like *gzip* normally have just two phases, an encoding phase and a decoding phase. Our model adds a third phase, the schema induction phase. The basic assumption behind the induction phase is that it is possible to look at a small sample of training data and infer a schema that will provide a useful characterization of unseen records. We view records as coming down a pipe in a never-ending flow. This flow, or time dimension, leads to a natural asymmetry. The columns represent distinct processes, and the rows represent the outputs of these processes at different points in time. The table has a fixed width, but an ever increasing length (records per month). Under this view, the warehouse is just a big buffer that stores the most recent 1-100 months of this flow. The actual database could never be materialized. It spans forward and backward over all time. The problem for the induction process is to look at a small sample of this flow and make useful predictions. The inference process has to be causal. Transformations over the columns are relatively easy to apply to unseen data. We don't know how to make transforms over rows (time) that will generalize in quite the same way.

4   Conclusions

About 4 years ago AT&T Labs initiated an effort, called the AT&T InfoLab, aimed at researching technology and uses of the data that flows through a large telecommunications service company, doing it at full scale, and focusing on near real time application of information. There are, of course, a wide variety of activities associated with the movement, management, use and visualization of telecommunications data, as there are a wide variety of types of data within the telecom world. Figure 2 is an outline of the expertise necessary to address this agenda. It is our experience that expertise in all of these areas is a necessary condition to real progress.

AT&T InfoLab

(DATA => DECISION => ACTION) --- IN REAL TIME

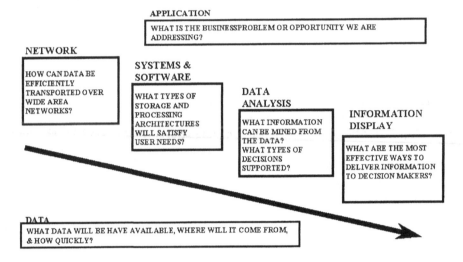

Fig. 2. Overview of the InfoLab

The InfoLab effort generally concentrates on data that is part of the telecommunications operations, administration, maintenance, and provisioning (OAM&P) processes for all of the major levels of telecom operations. That is the TNM levels of Network, Service and Business, along with the additional levels of Customer and Market. There is little concentration on Network Elements or Element Managers.

As part of a research effort in the early 1990s, we discovered from a review of the Architecture Reviews of several hundred OAM&P systems, that these systems are combinations of only three different types of data architectures:

- Transaction Processing (databases of record)
- Decision Support (data warehouses)
- Dataflow (transport).

We further discovered that such systems can effectively be partitioned into interoperating subsystems of single architectural types (Belanger et al., 1996). Each of these architectural types has a number of well-known examples. Examples of transaction databases in telecommunications operations include: network inventory, facilities (e.g. trunks), and customer databases. These are often large, and especially in the case of the facilities databases, extremely volatile.

The decision support databases, often called data warehouses, are likely combinations of the transactional bases and event related activity. For example, a marketing decision support database could contain much of the customer information data and the usage data for that customer. We see databases in the range of several tens of terabytes in these applications. Update is often batch, and the gating problem is often the load speed of the database management system as opposed to query speed.

Finally, the dataflow systems are non-traditional databases in the sense that much of their value, perhaps most of it, is realized prior to permanent storage. In simple architectural terms, they are pipes and filters followed by storage, rather than files with queries. They are the source of the most voluminous of the OAM&P data sets that we see. They derive from virtually any type of event on any of our networks. Most such events are results of usage or monitoring of the network (or the services that run in it). For example: telephone calls (call detail); packet headers (for IP, frame relay, or ATM traffic); logs, alerts or alarms (signaling data). The identifying characteristic of this category is that it can be treated as a sequence of records that can be accessed as a near real time flow. Our volumes are on the order of 300 million call detail records (about 25 gigabytes) per day, a similar volume of packet data, and a lesser amount of network signaling data. Each of these volumes will increase by integral factors in the next few years due to growth in local and data communications traffic. The classic applications of this form of data include network management and operations, fraud detection, and billing.

In this note, we have illustrated several uses of dataflow architectures with issues, results and example applications. Dataflows are very important in the telecommunication business, which is, after all, very much about the transport of data across networks. Within the InfoLab, we have built a dozen or more prototype systems investigating related architectures, along with a number of tools to support the systems. Most of these systems are now used in production, often with additional development beyond the prototype stage. This note mentioned two applications briefly, fraud and marketing, and two others in more detail, *Gecko* and *pzip* (compressing a call detail data warehouse). The discussion focussed on how dataflow architectures are being used in telecom OAM&P applications.

Acknowledgements. Many people have contributed to the AT&T InfoLab. Scott Daniels, Andrew Hume, Jon Hunt, Angus Maclellan and many others are responsible for the success of Gecko. Don Caldwell, Kenneth Church and Glenn Fowler developed *pin* and *pzip*.

References

1. Belanger, D.G., Chen, Y.F., Fildes, N.R., Krishnamurthy, B., Rank, P.H., Vo, K.P., Walker, T. E., "Architecture Styles and Services: An Experiment Involving the Signal Operations Platforms - Provisioning Operations System," *AT&T Technical Journal*, V75 N1, Jan/Feb 1996.
2. Bell, Whitten, I., and Cleary, J., *Text Compression*, Prentice Hall, 1990. Cortes, C., and Pregibon, D., "Giga-Mining, *Proceedings of the Fourth International Conference on Knowledge Discovery and Data Mining*, 1998.

3. Cortes, C., and Pregibon, D., "Information Mining Platforms: An infrastructure for KDD rapid deployment, *Proceedings of the Fifth International Conference on Knowledge Discovery and Data Mining*, 1999.
4. Goldstein, J., Ramakrishnan, R., and Shaft, U., "Compressing Relations and Indexes," University of Wisconsin-Madison, Technical Report No. 1366, 1997.
5. Gray, J. and Reuter, A, *Transaction Processing: Concepts and Technology*, Morgan Kaufman Publishers, San Mateo, Calif, 1993
6. Hernandez, M. and Stolfo, S., "The Merge/Purge Problem for Large Databases," *SIGMOD*, pp. 127-138, 1995.
7. Hume, A. and Maclellan, A., "Project Gecko: pushing the envelope," *NordU'99 Proceedings*, 1999.
8. Nelson, M., and Gailly, J., *The Data Compression Book*, M&T Books, New York, NY 1995.
9. Ray, G., Haritsa, J. and Seshadri, S., "Database compression – A perfromance enhancement tool," *VLDB*, 1995.

Telecommunications Databases –
Applications and Performance Analysis

Matthias Jarke and Matthias Nicola

Technical University of Aachen, Informatik V (Information Systems)
Ahornstr. 55, 52056 Aachen, Germany
{jarke,nicola}@informatik.rwth-aachen.de

Abstract. This paper first presents a short overview of application areas
for database technology in telecommunications. The main result of this
brief survey is that telecommunications has recently broadened to a degree that almost all fashionable fields of database research are becoming
relevant; nevertheless, relatively little research focus has been set on this
application domain. In the second part of the paper, we look specifically
at the performance challenges posed to distributed and wireless database
technologies by novel applications such as Intelligent Networks, mobile
telephony and very large numbers of database caches embedded in telephones. A brief overview of two of our own research efforts is presented,
one concerning the design of replication mechanisms, the other concerning analytic performance models in distributed and wireless information
systems.

1 Introduction

Due to technological advances and deregulation, the telecommunication market
is becoming increasingly competitive. The rapid evolution of the telecommunication business and the pressure of competition leads to a change in the telecommunication market. Traditionally, Telco has been a very product oriented mass
market in which one or few products and services aimed at millions of "identical"
customers. With growing competition customer satisfaction is becoming extremely important to stay in the marketplace. Therefore, the telecommunication
market has evolved towards customer oriented business in which products and
services are tailored to the individual needs of the end users. This change entails
a variety of data management and database challenges in the Telco business. Customer relationship management, market analysis, the evaluation of call detail
records, analysis of customer churn, complex billing systems and personalized
telecommunication services require very efficient database support.

2 Databases in Telco: Research and Vendor Effort

How much is the database research community and how much are the database
vendors already involved in the telecommunication business and its specific database needs? To get a first and rough idea we conducted a quick Internet based
survey.

W. Jonker (Ed.): Databases in Telecommunications, LNCS 1819, pp. 1–15, 2000.

In the first part of the survey, the aim was to identify research papers which deal with databases in telecommunications and have been published in leading database journals or proceedings from 1995 – 1999. Therefore we performed a simple title-search in Michael Ley's DBLP, the digital bibliography on database systems and logic programming (`http://dblp.uni-trier.de/`). We looked for papers containing words like "telecommunication", "telco", "telephony", "phone", etc. in their title. The resulting set of papers which fit to the scope of this workshop is given in Table 1. Despite the rough and inaccurate nature of this survey it is becomes apparent that the database research community has neglected telecommunications as a very important field of database applications.

Table 1. Telecom-Papers in the Database Literature 1995-1999

Conference / Journal	No.	Reference
EDBT	1	[3]
ICDE	1	[12]
SIGMOD	1	[18]
VLDB	2	[14], [1]
TODS	0	
TKDE	0	
Distributed and Parallel Databases	1	[2]

In the second part of the survey, we wanted to get an idea whether or not major database vendors have paid more attention to the growing Telco field. As a rough criterion we looked at the success stories provided at the companies' web sites. The result in Table 2 shows that telecommunication has been recognized by the database vendors but is not as big as a field of application as traditional domains such as banking or other growing fields like media and entertainment.

3   Database Applications in Telecommunications

Looking at the success stories described, database applications in telecommunications can be classified as sketched in Figure 1. The range of conventional *OLTP databases* covers mainly billing and call routing tasks as well as location management in mobile phone systems. A second area of database application in telco is *knowledge management*. On the one hand, knowledge management includes the growing field of data warehousing and data mining. Typical applications are market and marketing analysis, churn analysis, fraud detection, customer relation management (CRM), and the evaluation of call detail records (CDRs) and other traffic characteristics. On the other hand, knowledge management also subsumes repositories for telecommunication software, e.g. for service

Table 2. Number of success stories by industry and vendor

	IBM	Oracle	Sybase
Banking / Finances	138	11	29
Health Care	92	4	22
Media / Entertainment	4	12	21
Telecommunications	0	12 (??)	10

creation and execution in Intelligent Networks or re-engineering of legacy software. While OLTP and knowledge management applications are employed to operate telecommunication systems and businesses, *Information services* describes a third field of database applications in which telecommunication enables database applications and –to a certain extent— vica versa. The advances in mobile communications has lead to the development of lightweight database systems for database functionality anywhere anytime.

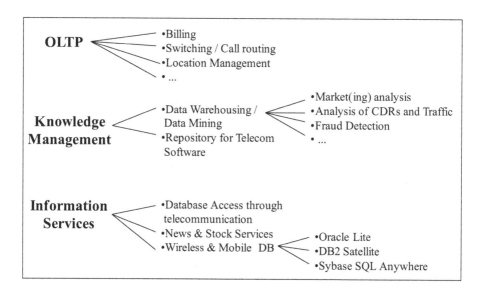

Fig. 1. Database stakes in telecommunication

4   Managing Telecom Data with ADR

The general idea of ADR (Atomic Delayed Replication) is to combine replication management with distributed database design. ADR provides means to define and alter the database and the replication schema, as well as mechanisms for atomic but asynchronous and possibly delayed propagation of updates. In sections 4.1 and 4.2 we describe how ADR is based on an application oriented partitioning of data so that different levels of consistency can be defined and maintained.

ot all database approaches to telecommunications management problems are using distributed database technology at all. For example, the ClustRa main memory parallel database [14] is based on a shared-nothing approach, ATM inter-node communication, and hot stand-by secondary copies to achieve high availability. A major difference to the ADR system is that ClustRa is expensively being built up from scratch while ADR is built on top of inexpensive commercial database technology. Other central main memory oriented telecom databases include the *Dali* system from AT&T Bell Labs [15], Hewlett-Packard's *Smallbase* [13] and Nokia's *TDMS* [17]. These databases are highly specialized to specific telecommunication systems while ADR is sufficiently flexible to support a variety of applications.

4.1   Data Partitioning

Replication is based on the primary copy approach where replicas (primary and secondary copies) are defined on the granularity of so-called *partitions*. Partitioning requires an additional database design step named *grouping* as an intermediate step between fragmentation and allocation, as depicted in figure 2. Data fragments which are logically closely related either because of integrity constraints or because of frequent joint usage are grouped together to partitions which represent the units of allocation, and thus replication.

The aim is to define partitions such that most consistency requirements are partition *internal*. Transactions are defined to consist of n read-steps and at most one write step such that different steps access logically independent data items. A skilful partition schema should then allow a major share of the transaction steps to be executed on single partitions. Restricting transactions to a maximum of *one* write step is a crucial requirement for the consistency properties guaranteed by ADR.

The example below shows a (slightly simplified) transaction related to a phone call using a virtual private network (VPN) service. The transaction splits into a write-step and a read-step. In the write-step, a counter variable (CNT) for the service 600 of subscriber 1234 is increased for statistics and billing. The read-step is used to find the real world telephone number assigned to the short number SN_11 that has been used to initiate the call. Obviously, the operation of increasing the counter is independent from the value of the phone number read. Still, both steps have to be within the context of a single transaction to provide atomicity: an abort during the read-step also requires to undo the write-step because we do not want to charge users for calls which could not get connected.

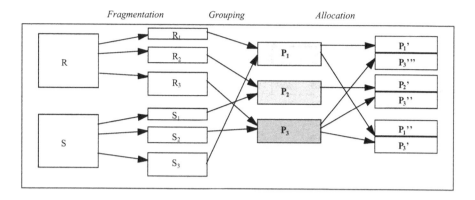

Fig. 2. Partitioning as a database design step

```
BEGIN TRANSACTION

  BEGIN WRITE STEP 'S.600.1234'
    UPDATE variable SET value = value + 1
      WHERE serv_nr = 600 AND subs_nr = 1234 AND var_name = CNT

  BEGIN READ STEP 'U.600.1234'
    SELECT value FROM variable
      WHERE serv_nr = 600 AND subs_nr = 1234 AND var_name = SN_11

COMMIT TRANSACTION
```

Figure 3 shows the partitions used by the sample transaction. The operational part of the VPN service owned by subscriber 1234 is defined by a single record in the table *Subscriber* and by the VPN numbers (SN_11, SN_12 and SN_13) in the table *Variable*. These records are grouped to a partition "U.600.1234". The statistical part of the same VPN consists of the counter record in the table *Variable* and a description record in the table *Service* which form another partition named "S.600.1234". The partition "U.600.1234" will rarely be changed but read very often. Therefore it is clever to replicate it. Partition "S.600.1234" should not be replicated because it is often modified.

Structuring partitions and transactions this way is surely not possible for every imaginable database application. However in many OLTP applications with short transactions over few records, the ADR idea can be employed. As the example indicates, we initially implemented and tested ADR in a distributed database system by Philips supporting *Intelligent Networks* (IN).

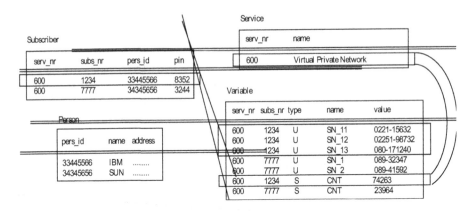

Fig. 3. Partitions for a VPN service in the intelligent network

4.2   Internal and External Consistency

Partitions are replicated according to the primary copy approach, i.e. there is one primary copy and m secondary copies. This approach is followed by most replication algorithms; its theoretical merits have been recently argued by [8]. Updates are propagated from primary to secondary copies asynchronously, i.e. not within the context of the original update transaction and possibly delayed. Hence, secondary copies may age but still provide a sufficient level of *partition internal consistency* that the application is satisfied with.

Definition: A partition is called *internally consistent* if all partition internal consistency requirements are fulfilled.

Definition: A set of partitions are called *externally consistent* if all (internal and global) consistency requirements are fulfilled.

The ADR system ensures that the set of primary copies is always externally consistent; secondary copies are always internally consistent but may be out of date. Furthermore, a transaction's write-step (if any) always has to be executed on the primary copy, perceiving (and preserving) external consistency. Read-steps can be carried out on any secondary copy, as long as the application is satisfied with internal consistency. Otherwise read operations have to be included in a write-step. For instance, imagine a transaction that consists of two read steps reading two different secondary copies SC1 and SC2. ADR guarantees that both read steps see a state of SC1 and SC2 respectively, which once was a valid state of the respective primary copy. Yet, the states of SC1 and SC2 may be of a different age such that SC1 and SC2 may reflect a combination of values which never existed among the related primary copies. If consistency regarding *both* partitions (*partition external consistency*) is required, then the two reads have to be embedded in a write step. This forces the read operations to be executed on the primary copy.

Since every transaction has at most one write-step, atomicity of an update transaction can be ensured by a single site, i.e. the site holding the primary copy. ADR does not require distributed concurrency control but the possibly distributed read-steps can be executed under local concurrency control at the speed of a centralized DBMS. The synchronous two phase commit protocol is only used in the rare case that a write step needs to access multiple partitions which are located at two or more different sites.

4.3   Overall System Architecture

ADR has been implemented on top of commercial relational database technology, namely Sybase SQL Server. The general system design is depicted in Figure 4. Applications access databases through the ADR module. To avoid hindering the database's communication parallelism to the application site, and to minimize the communication overhead between the application and ADR, ADR is not running at the database site but at the application sites where it is a software library linked to the application source code.

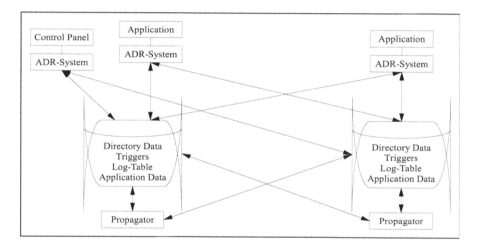

Fig. 4. Overall system architecture

The databases hold the application data as well as meta-data, triggers and log-tables which are used to perform transaction processing and replica management according to the ADR formalism. So-called propagators at each database site are in charge of executing reproduction transactions correctly; they are independent from the application and its data. (Reproduction transaction are used to update secondary copies). The database administrator can use a control panel to change the replication schema (e.g. number and placement of secondary copies) or even to extend the relational database schema, without interrupting the current database activities or application programs.

5 Applications and Performance

ADR has been used in two real-world applications to allow an authentic and meaningful evaluation. One is a distributed database for *Intelligent Network* (IN) telephone services; the other is real-time data support for mobile phones in a city-wide DECT setting. The main performance goals of the IN application are high throughput and scalability while providing sufficiently low response times. In the city-wide DECT application the most critical performance requirement is extremely short response times. Experiences with these two applications have shown that ADR is indeed suitable to allow for high throughput and scalability or for very short response times respectively by relaxing coherency in a controlled manner. Sections 5.1 and 5.2 present the experiences with ADR in the Intelligent Network and city-wide DECT application respectively.

5.1 Integrating Database Design and Operation for Intelligent Network Services

This part of our research was initiated by the need to provide database support for an Intelligent Network design and operations environment developed by Philips Labratories. The IN is an architectural concept for telecommunication networks that enables network operators as well as independent service providers to swiftly introduce new services such as free-phone, virtual private network, televoting, etc. into existing networks. Furthermore, these services should be made sufficiently flexible so that after deployment, service subscribers can tailor them to their requirements.

The main idea of the IN concept is the separation of switching functionality from service control. To achieve a high degree of flexibility, the service logic is realized by software modules called *service logic programs* (SLPs) which can be customized with subscriber specific data. Figure 5 shows the structure of the Intelligent Network as it is defined by the IN standards ITU-T (former CCITT) CS1 and AIN.

A *Service Switching Point* (SSP) recognizes calls from an end user phone to a service which requires support by a *Service Control Point* (SCP) and sends an instruction request to the SCP. A SCP retrieves the corresponding *Service Logic Program* (SLP) and service data from the database, evaluates it and sends a response back to the SSP. The *Service Creation Environment* (SCE) is used for creation and testing of new services which are then transferred via the SMS to the SCP. The *Service Management System* (SMS) is needed for downloading service logic programs and service data as well as for other management activities such as billing and statistics.

Such an IN system has to handle large amounts of data (SLPs, subscriber specific data, management information). Several IN vendors are using a large central mainframe database system to provide consistent data support for all SLPs running at the same time. However, such systems are not only very expensive but also a potential bottleneck regarding availability and scalability. Furthermore, every data request requires communication with the central site.

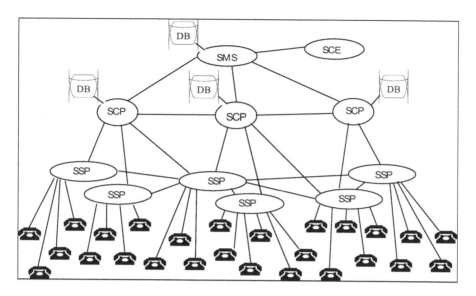

Fig. 5. The Intelligent Network Architecture

As telecommunication systems are of highly parallel nature, a large scale distributed database system composed of commodity hardware can be a more natural and less expensive solution [7]. Hence, a main design goal of our ADR-based implementation was to provide high performance and highly scaleable replica management on top of standard database hardware and software. The implementation of ADR in the IN database followed the outline in section 4 and is elaborated in [5],[6].

Using our queuing model from [4], one of the most important performance results for the IN application is given in Figure 6. It shows the maximum throughput as a function of the number of sites n. The coherency index is taking the values 0, 0.1, 0.25, 0.5, 0.75 and 1 (from top to bottom), where $k = 1$ stands for immediate propagation of updates to secondary copies while $k = 0$ represents unbounded divergence between primary and secondary copies. If the percentage of updates is not negligible (like 10% in Figure 6), throughput does not increase linearly with the number of sites due to update propagation overhead (when $k > 0$). However, the graphs for $k < 1$ indicate that relaxed coherency may improve scalability towards ideal linearity. Figure 6 also shows that for a given number of sites throughput can be increased by relaxing coherency, and the larger the system the greater the gain.

The analytical model has been validated through measurements of ADR in the IN implementation (section 5 in [4]). It has shown that ADR is a very suitable concept for the transaction processing and replica management which is required to provide distributed database support for an intelligent network. In particular, the coherency trade-offs in ADR allow for sufficient throughput and scalability as demanded by a large scale distributed telecommunication system.

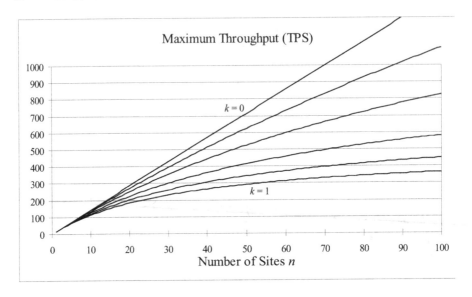

Fig. 6. Maximum throughput vs. system size for 90% read-only transactions

5.2   Managing Mobility Data in the City-Wide DECT

The *City-Wide DECT* is a mobile, wireless telephone network using very small cells[1]. In the Philips implementation, so called Mobility Managers (MM) provide the system with profile and location information about the users. This information has to be administrated in a distributed (replicated) database due to the distributed nature of the overall system. As for the intelligent network, we believe that the MM components and their databases should be realized on commodity hardware available in the mass market for cost reasons and independence of a special hardware manufacturer.

In our implementation, a reference copy of the complete information (location and profile) is placed as a primary copy in a conventional relational database management system. This sub-component provides reliability by means of classical storage on persistent media (hard disks). In order to achieve high availability of the primary data we used concepts directly supported by the RDMBS system, like RAID (redundant arrays of independent disks). For recovery reasons, the data and the log of the database were placed on separate disks. Furthermore, the RDBMS directly supports the mirroring of data-disks and log-disks. Our experiences show that these high availability concepts did not slow down the RDBMS machine. In order to increase availability we introduced multiple hot-standby machines. The redundant secondary copies are driven through ADR concepts, i.e. relaxed consistency and asynchronous propagation of updates.

In contrast to the IN application, the most critical performance requirement of the City-Wide DECT application is a very short *response time*. A mobile

---
[1] DECT stands for Digital Enhanced Cordless Telecommunications.

phone user expects a dial tone less than one second after she lifts the receiver. Analyzing the tasks to be performed during this second (i.e. "attach" to the network) led to the result that at most 10 ms can be spared for database access. The performance evaluation of the IN application indicated that such a response time can hardly be achieved with today's commodity database and disk technology. Indeed, additional measurements proved that a single conventional centralized database server is not able to fulfill such a tough response time requirement [6].

Interestingly, the ADR approach works equally well if secondary copies are kept in main memory databases or caches; such caches were therefore placed in each MM. This allows to access the location data without disk I/O or remote access to the primary copy. The most difficult problem that remained was to propagate the location updates that occur in a distributed manner. We relaxed the coherency of the secondary copy in a time oriented manner through delayed propagation of updates. The City-Wide DECT is meant to provide seamless service for users at up to walking speed; thus, location information can be tolerated to age up to 10 seconds due to overlapping cells. Thus, updates at the primary copy are collected for at most 10 seconds and then written to the secondary copies in a single transaction. As shown below, our analytical model predicted that this will achieve a response time below 10 ms while guaranteeing that the MM process never accesses data older than 10 seconds. Extensive measurements on the implementation confirmed these results.

The City-Wide DECT application can be represented in queuing model from [4] by setting the number of sites n to 10 and the overall transaction arrival rate to 100 TPS, where only 50% are assumed to be read-only transactions, because decreasing cell sizes and high traffic rates will lead to a very high update rate for location information in mobile telephone networks [16]. This means the distributed system has to execute 50 updates per second leading to 5 updates per seconds per site. Thus, within the 10 seconds of update propagation delay a number of 50 updates are accumulated into 1 update transaction which is forwarded to the secondary copies. This yields a coherency index of $k = 1/50 = 0.02$.

Figure 7 shows the expected average transaction response time as a function of the degree of replication for different levels of coherency requirements. The case $k = 1$ represents asynchronous but immediate propagation of updates. In this situation the response time can be reduced remarkably by replicating about 30% of the data. This leads to increased local access, while a higher degree of replication rapidly saturates the local databases with propagated updates ($r > 0.4$). The graph for $k = 1$ shows that even with an optimal degree of replication asynchronous but immediate update propagation would prevent the database from satisfying the response time requirement so that a relaxation of coherency is necessary. The curve for $k = 0.5$ represents the case of refreshing the secondary copies after every second update and $k = 0.1$ means to delay update propagation for intervals of 2 seconds. However, neither strategy is suitable to decrease response time below 10 ms. The allowed delay of 10 seconds ($k = 0.02$) has to be fully exploited to reduce the load of processing reproduction transactions far enough such that full replication becomes affordable and response time drops below 10 ms. In order to verify this result we carried out measurements in our ADR implementation for the City-Wide DECT application. We generated about a million typical City-Wide DECT transactions and found that in the case of full replication and 10 seconds update propagation delay, indeed 99% of the transactions had a response time of less than 10 ms.

6 Towards Improved Performance Models

The experiences made in the ADR project have been generalised in a comprehensive set of analytic performance models for distributed and replicated databases, with a focus on wireless information systems [10]. The basic observation underlying this work is that existing performance models of distributed and replicated databases can only represent relatively simplistic replication schemes and hardly consider the interplay between database and communications aspects of distributed, replicated and possibly wireless information systems. Figure 8 shows what, in our opinion, should be captured by realistic models.

In [11], we present a critical survey of about 40 proposals in the literature according to these criteria and propose a generalised scheme called 2RC (2.dimensional replication with communication) which enables analytical modeling of the interplay in figure 8. The dependencies considered in this model are shown in figure 9. The figure demonstrates that the model considers multiple transaction types both from a replication and a communication perspective; moreover, replication is modeled both in terms of data items replicated and number of replicas produced. These properties may seem obvious and still too simple but have interestingly not been reached by previous models. Specifically, they enable modeling the quality of replication as well as bottleneck analysis in settings where both the database and the network can become the bottleneck.

However, for realistic telecommunications database applications, more aspects have often to be considered. In [9], the 2RC approach has therefore been extended to capture specific aspects of database technology, such as locking conflicts, and specific aspects of wireless communication technology, such as broadcast networking, delays for connection setup and propagation, unreliable and unavailable communication links. It is shown that analytical performance models are still feasible as long as these extensions appear in isolation, but that certain combinations of these extensions lead to models that can only be solved by simulation.

The extended 2RC models have been validated by system measurements using an extension of the DebitCredit Benchmark. Moreover, two real-world applications in the telecom sector have been studied. The first one, conducted in cooperation with Ericsson Eurolabs in Aachen, considers the design of a wireless information base through which users of GSM-based mobile phones can obtain efficient access to database information such as news or stock data; in practice, this amounts to the question how a scalable architecture for millions of mobile database caches can be reached. The second application is at the other end of

Response time in seconds

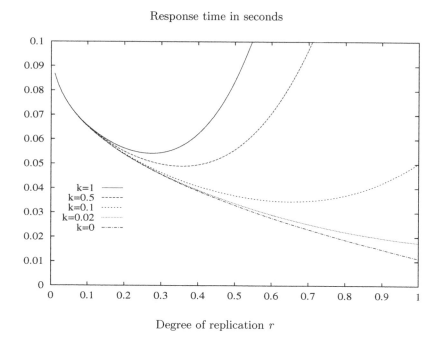

Degree of replication r

Fig. 7. Response time results for the City Wide DECT application

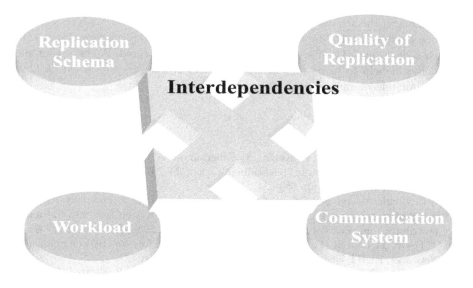

Fig. 8. Desiderata for performance modelling in distributed replicated databases

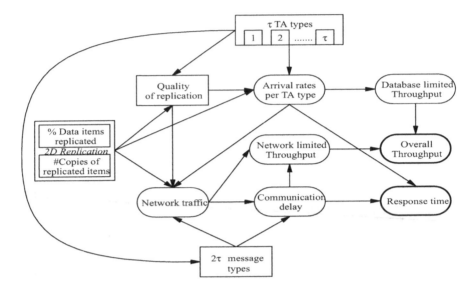

Fig. 9. Captured dependencies in the 2RC performance modeling approach

the technological spectrum. It considers the design of a low-cost wireless information system for networks of mission hospitals, local health centers and simple dispensaries in rural Tanzania, using packet radio technology and cheap standard PC databases. A design based on the models could be identified, and the corresponding system has been implemented with good initial success in terms of improved information flow about mobile patients, statistical information, and epidemiological surveys.

In summary, the results seem to support the practicality of analytical performance models for distributed, replicated, and wireless information systems in a much wider range of applications than previously considered. We hope that such models will help to bring forward the frontiers of scientifically based database design for telecommunications.

References

1. J. Baulier, S. Blott, H. F. Korth, A. Silberschatz: A Database System for Real-Time Event Aggregation in Telecommunication. 24th International Conference on Very Large Databases, pp. 680-684, (1998)
2. A.B. Bondi, V. Jin: A performance model of a design for a minimally replicated distributed database for database driven telecommunication services, Journal on Distributed and Parallel Databases, Vol. 4, No. 4, pp. 295-317, October (1996)
3. J. Fessy, Y. Lepetit, P. Pucheral: Object Query Services for Telecommunication Networks, EDBT '96, pp. 490-493, (1996)
4. R. Gallersdörfer, M. Nicola: Improving Performance in Replicated Databases through Relaxed Coherency, Proceedings of the 21th International Conference on Very Large Database, pp. 445-456, September (1995)

5. R. Gallersdörfer, K. Klabunde, A. Stolz, M. Eßmajor: Intelligent Networks as a Data Intensive Appliction - Final Project Report, Technical Report AIB-96-14, ISSN 0935-3232, Technical University of Aachen, June (1996)
6. R. Gallersdörfer, M. Jarke, M. Nicola: The ADR Replication Manager, International Journal of Cooperative Information Systems (IJCS), Vol. 8, No. 1, pp. 15-45, March (1999)
7. D. J. DeWitt, J. Gray: Parallel Database Systems: The Future of High Performance Database Systems. CACM 35(6): 85-98 (1992)
8. J. Gray, P. Helland, P. O'Neil, D. Shasha: The Dangers of Replication and a Solution. SIGMOD Conf. 1996: 173-182 (1996)
9. M. Nicola: Performance Evaluation of Distributed, Replicated, and Wireless Information Systems Doctorate Thesis, Technical University of Aachen, Department Informatik V, AIB-99-10, October (1999)
10. M. Nicola, M. Jarke: Increasing the expressiveness of analytical performance models for replicated data. Proc. ICDT 99, Jerusalem, January (1999)
11. M. Nicola, M. Jarke: Performance modeling for distributed and replicated databases. Research survey, conditionally accepted for publication in IEEE Transactions on Knowledge and Data Engineering, (2000)
12. K. Hätönen, M. Klemettinen, H. Mannila, P. Ronkainen, H. Toivonen: Knowledge Discovery from Telecommunication Network Alarm Databases, ICDE '96, pp. 115-122, (1996)
13. M. Heytens, S. Listgarten, M.A. Neimat, K. Wilkinson: Smallbase: A main memory DBMS for high performance applications, Hewlett-Packard Laboratories, March (1994)
14. S.-O. Hvasshovd, Ø. Torbjørnsen, S. Erik Bratsberg, P.r Holager: The ClustRa Telecom Database: High Availability, High Throughput, and Real-Time Response, 21st International Conference on Very Large Databases, pp. 469-477, (1995)
15. H.V. Jagadish, D. Lieuwen, R. Rastogi, Avi Silberschatz: Dali: A high performance main memory storage manager, Proceedings of the 20th International Conference on Very Large Databases, pp. 48-59, September (1994)
16. C. N. Lo, R. S. Wolff: Estimated Network Database Transaction Volume to Support Wireless Personal Data Communications Applications, Proceedings ICC '93, Genf, May (1993)
17. M. Tikkanen: Objects in a telecommunications oriented database, Proceedings of the Conceptual Modelling and Object-Oriented Programming Symposium, (1993)
18. S. Trisolini, M. Lenzerini, D. Nardi: Data Integration and Warehousing in Telecom Italia, SIGMOD Conference, pp. 538-539, (1999)

Joining Very Large Data Sets

Theodore Johnson[1] and Damianos Chatziantoniou[2]

[1] Database Research Center, AT&T Labs – Research
johnsont@research.att.com
[2] Dept. of Computer Science, Stevens Institute of Technology
damianos@cs.stevens-tech.edu

Abstract. Many processes in telecommunications (e.g., network monitoring) generate very large amounts (many terabytes) of data. This data is stored in a data warehouse and used for data mining and analysis. Many analyses require the join of several very large data sets. Conventional methods for performing these joins are prohibitively expensive. However, one can often exploit the temporal nature of the data and the join conditions to obtain fast algorithms that operate entirely in memory. In this paper, we describe such a join algorithm (the *window join*) together with a method for analyzing queries to determine when and how the window join should be applied. The window join makes sequential scans over the input data, allowing the use of tape storage. We have used the techniques described in this paper on a large IP data warehouse.

1 Introduction

The rapidly declining cost of digital data storage has encouraged the creation of *data warehouses*. In a typical scenario, an organization stores a detailed record of its operations in a database, which is then analyzed to improve efficiency, detect sales opportunities, and so on. In many applications, the warehoused data sets can become very large. For example, AT&T creates data warehouses of network operations data. This information is vital for optimizing network operations, debugging network problems, fraud detection, verification of billing procedures, and so on. These data sets are collected at the rate of tens of gigabytes per day, and rapidly accumulate into multiple terabytes.

The scale of these very large data sets makes it difficult to analyze them using conventional database technology. Analysts often need to evaluate complex aggregates on the data sets. When expressed in SQL, the queries often involve very large joins and self-joins, creating temporary views, and joins between the views and the data sets. We have found that commercial DBMSs cannot optimize these queries properly, so their evaluation is prohibitively expensive. As a result, these queries are usually written in a procedural language (e.g., PL/SQL, Perl, etc.) that reads data from the DBMS or from a flat file export from the database. However, this is not an entirely satisfactory solution because these queries must usually be written by a skilled database administrator, and their time is a very scarce resource.

W. Jonker (Ed.): Databases in Telecommunications, LNCS 1819, pp. 118–132, 2000.
© Springer-Verlag Berlin Heidelberg 2000

In addition, storing these data sets on-line can be prohibitively expensive in spite of the rapid decline in the cost of magnetic disk drives. Tape-resident tertiary storage can provide a high-performance and low-cost storage alternative. A typical installation uses an automated tape library, which consists of a storage rack for a collection of tapes, a set of tape drives, and a robot arm that transfers tapes between the storage rack and the drives. Tape-resident tertiary storage is one to two orders of magnitude less expensive than on-line storage on a per-byte basis [10, 6]. However, tape drives are inherently sequential [9]. While tape resident data can be made available to a DBMS, it can only be sequential scanned.

These considerations have motivated our research into query languages and evaluation algorithms suitable for very large scale aggregation. We have found that the EMF SQL language [3, 4, 5, 8] to be well suited to our needs, as it is designed to express complex aggregation. In a previous paper [4], we have described how a very large scale data warehouse can be stored primarily on tape and still efficiently process complex aggregation queries. This paper discusses aggregation over a single table, and assumes that all of the dimension tables that are joined to the fact table are small. However, many queries require that two (or more) very large tables be joined.

In this paper, we present a method for joining two very large relations. Our technique makes use of the temporal nature of the data sets commonly encountered. In some ways, our technique is similar to the diag-join proposed in [7]. Both techniques rely on the joined tables being sorted by time, and on the joined tuples being temporally close to each other. However, there are two significant differences between our method and the diag-join:

1. The diag-join is an equijoin and requires that most joined tuples be close to their a-priori expected location for good performance. Our method is more similar to a join of a fact table to a very large temporal dimension table. The bounds on joined tuple locality is determined by an analysis of the query.
2. The diag-join produces an output relation, on which additional processing is required for aggregation queries with group-bys. Our method is designed to work in conjunction with an efficient large scale grouping and aggregation operator, although it can also produce the joined relation as output.

The query analysis and processing that underlies the window join derives from our work on very large scale aggregation in EMF SQL [4], so we present our join method in the context of EMF SQL. However, the query analysis can be abstracted and the window join used in conventional SQL. In Section 2, we discuss an example (single table) query that is typical in our application. In Section 3, we review EMF SQL. In Section 4 we present the query analysis required for the window join. We show how to use the EMF SQL query analysis techniques for the window join in Section 5. In Section 7 we conclude.

2 Complex Aggregation on Telecommunications Data

The application that has motivated our research is the analysis of network performance data. A common feature of the problems encountered is the aggregation of sequences of tuples into larger units. In this section we give a (highly simplified) example.

Let us consider a scenario in which the data warehouse contains records describing packets that traverse a communications network. This relation has the following schema:

Pkt (Source, Dest, StartCall, EndCall, Length, ts)

Example 1: Suppose that a network engineer wants to know how many packets are involved in each connection, where a connection is a sequence of packets (in ts order) with the same Source and Dest attribute values, starting with a packet where StartCall is 1, and ending with a packet where EndCall is 1. Packets of different connections may be interleaved. We can assume that every packet has a unique value of ts. A direct expression of this query in SQL is

```
create view Temp as
select R.Source, R.Dest, R.ts, EndTime=min(S.ts)
from Pkt R, Pkt S
where R.Source=S.Source AND R.Des=S.Dest AND R.ts ≤ S.ts AND
      R.StartCall=1 AND S.EndCall=1
group by R.Source, R.Dest, R.ts

select T.Source, T.Dest, T.ts, count(*)
from Pkt R, Temp T
where R.Source = T.Source AND R.Dest=T.Dest AND
      R.ts ≥ T.ts AND R.ts ≤ T.EndTime
group by T.Source, T.Dest, T.ts
```

This query is expressed as one cross-product self-join, the creation of a temporary table, and another join. This type of query is difficult to optimize, and in fact our experience is that common commercial SQL databases cannot optimize this type of query properly.

However, when a programmer is asked to compute the number of packets per connection, she will usually write an efficient program that scans the Pkt table and builds a hash table of aggregates indexed by the Source, Dest pairs. It is clear that a packet with StartCall=1 defines a new hash table entry, and a packet with EndCall=1 closes the entry. If we increment a counter for every packet that hashes to an entry, when we close the entry we have computed the output. Finally, closed entries can be deleted from the hash table to minimize space use.   □

As the example shows, there is a large class of queries that are difficult to express and optimize in SQL, but which have simple and natural programs for computing their result. The motivation behind EMF SQL is to express these queries in a declarative way.

3   The Extended Multi-feature Syntax (EMF SQL)

The development of the window join is based on concepts developed for EMF SQL query processing. In this section, we briefly review EMF SQL syntax, semantics, and processing. For a fuller treatment, please see [3].

The idea behind the extended multifeature syntax [2] is simple. For each group, the user defines one or more *grouping variables*. Each grouping variable represents a subset of the entire relation, whose range is constrained by the *such that* clause. The defining condition of a grouping variable may contain comparisons between ordinary attributes and constants, aggregates of the group, and aggregates of *previously defined* grouping variables. As a result, one may define a series of selections and aggregations over the same grouping attributes. The group itself can be considered as one grouping variable, denoted as X_0. Aggregates of the grouping variables can appear in the select clause.

This small extension to SQL allows the user to express nearly all important decision support queries in a simple and declarative fashion. This is mainly achieved because the group by clause acts as an implicit iterator over groups, the same way the from clause acts for the tuples of a relation. At the same time, grouping variables define the processing to be done for each value of the grouping attributes.

3.1   Syntax and Semantics

The extended multi-feature syntax modifies standard SQL in the following ways:

- **From, Where** clauses: There are no changes.
- **Group by** clause: The declaration of the grouping variables follows the grouping attributes, separated by a semicolon:

$$\text{group by } G_1, \ldots, G_m \; ; \; X_1, \ldots, X_n$$

- **Such that** clause: This newly introduced clause defines the range of the grouping variables. It has the following format:

$$\text{such that } C_1, \; C_2, \; \ldots, \; C_n$$

  where each C_i is a potentially complex expression (similar to a where clause), used to define X_i, $i = 1, 2, \ldots, n$. It involves comparisons between an attribute of X_i and either a constant, a group's aggregate, a previously defined grouping variable's aggregate, or one of the grouping attributes. Grouping attributes are considered constants in the such that clause (although they have different values in different groups).
- **Select** clause: It may contain the grouping attributes, aggregates of the group, and aggregates of the grouping variables.
- **Having** clause: It is an expression involving the grouping attributes, constants, the aggregates of the group, *and* the aggregates of the grouping variables.

Let us consider a couple of examples:

Example 2: Suppose that one wants compute for each product and for each month the moving average of sales for the last three months, the next months, and the average sales during the month:

```
select Product, Month, avg(X.Sales), avg(Y.Sales), avg(Z.sales)
from Sales
Where Year = 1998
Group By Product, Month ; X, Y, Z
Such That  (X.Product=Product and X.Month<Month and
                X.Month ≥ X.Month-3 ),
            (Y.Product=Product and Y.Month=Month),
            (Z.Product=Product and Z.Month>Month and
                Z.Month ≤ Z.Month+3 ),
Having sum(Y.sales) < sum(Z.sales)/2 AND
        sum(Y.sales)+sum(Z.sales) >= sum(Z.sales)/2
```

The `Sales` relation is grouped by `Product`, `Month`. For each product and month, X contains sales of that product for the previous three months, Z for the next three months, and Y the sales during that month.   □

Example 3: A data analyst wants to know for each customer and product, how many purchases of the product by the customer had quantity more than the average quantity sale of that product. This is an extended multi-feature query and can be written as:

```
select Customer, Product, count(Y.*) from Sales
group by Customer, Product ; X,Y
such that  (X.Product=Product),
            (Y.Customer=Customer and Y.Product=Product
            and Y.Quantity > avg(X.Quantity))
```

The `Sales` relation is grouped by `Customer`, `Product`. Then *for each* (customer,product) value (i.e. for each group) two sets are defined, X and Y, where X contains *all* the sales of the current product and Y contains all the purchases of that product by the current customer that exceeded the average quantity sale of the product (`Y.Quantity > avg(X.Quantity)`.)   □

3.2   Evaluation and Optimization of EMF Queries

In this section we present a direct implementation of extended multi-feature queries and optimizations of that implementation. All aggregate functions are presumed to be distributive (holistic aggregates are discussed in [8]). We start with two definitions.

Definition 31: *A grouping variable Y depends on grouping variable X if some aggregate value of X appears in the defining condition of Y. This is denoted as $Y \rightarrow X$. If the defining condition of a grouping variable Y contains aggregates of the group or if Y is not restricted to the group, then the group is denoted as a grouping variable X_0 and we write $Y \rightarrow X_0$. The directed acyclic graph that is formed from the grouping variables' interdependencies is called* emf-dependency *graph.* □

Definition 32: *The output of a grouping variable X, denoted as* outp(X) *is the set of the aggregates of X that appear in either the* such that *clause, the* having *clause, or the* select *clause.* □

3.3   Evaluation

Let H be a special table, called the *mf-structure* of an extended multi-feature query, with the following structure. Each row of H, called *entry*, corresponds to a group. The columns consist of the value of the grouping attributes, the aggregates of the group, and the aggregates of the grouping variables. Let $X_1, \ldots, X_n$ be the grouping variables of the query, ordered by a reverse topological sort of the emf-dependency graph. The following algorithm computes correctly all the entries of H.

Algorithm 31: *Evaluation of extended multi-feature queries:*
```
for  sc=0 to n {
      for  each tuple t on scan sc {
            for all entries of H, check if the defining condition of
            grouping variable Xsc is satisfied. If yes, update Xsc's
            aggregates appropriately. X0 denotes the group itself.
      }
}
```
□

   This algorithm performs $n+1$ scans of the base relation. On scan i it computes the aggregates of X_i grouping variable (X_0 denotes the group.) As a result, if X_j depends on X_i (i.e. if aggregates of X_i appear in the defining condition of X_j), this algorithm ensures that X_i's aggregates will have been calculated before the jth scan. Note also that given a tuple t on scan i, *all* entries of table H must be examined, since t may belong to grouping variable X_i of several groups, as in Example 2: a tuple t affects several groups with respect to grouping variable X, namely those that agree on product with t's product.

   Algorithm 31 represents an efficient, self-join free direct implementation of the extended multi-feature syntax. This type of query processing algorithm corresponds closely to the type of procedural program which would be written by hand. However, because the queries are specified declaratively a wide variety of optimizations can be automatically applied.

Indexing: The constraints of a grouping variable X in a such that clause constrain the groups that a tuple can affect when X is processed. For example, it is clear that in Example 2 the mf-structure should be indexed by Product, and perhaps also by Month.

Pass Reduction: The aggregates of some grouping variables can be computed together. For example, in Example 2, the aggregates of X, Y, and Z can be computed together because none of these grouping variables constrain any other (more precisely, none is reachable from the other in the emf-dependency graph). In Example 3, Y is constrained by an aggregate of X and therefore must be computed after the aggregates of X. We introduce the grouping variable X_0 to denote the pass that finds the groups. A grouping variable X can be processed with X_0 if X is not constrained by any aggregate, and X is restricted to tuples within the group. In Example 2, Y can be computed with X_0, but X and Z must be computed after X_0.

Partitioning: The mf-structure can be horizontally partitioned and processed in a partition-wise manner. We can use this property to handle very large mf-structure, by choosing the partition size to be the available memory size and performing all processing in-memory. We can implement parallel processing by distributing partitions among processors. When processing a grouping variable, a tuple will in general need to be processed at each of the partitions. However, we can use the constraints in the such that clause to limit the partitions that a tuple is sent to (i.e., as we do with indexing).

In [5], we describe a relational algeraic operator, the md-join, that implements Algorithm 31. The md-join allows the expression of additional optimization strategies.

Performance: As is reported in [2], we wrote a translator that generates C or PL/SQL code from an EMF query, and implements the above mentioned optimizations (except for parallel processing). We wrote SQL and EMF queries for several example queries, and generated PL/SQL and C programs from the EMF queries. We found that the C programs that we generated were two orders of magnitude faster than any of the commercial database systems that we tested (in spite of the high overhead, the PL/SQL programs were as fast as the SQL queries).

4   Optimizations for Sequence Aggregates

In [4], we present optimizations that enable very large scale aggregation over sequence data. In this paper, we use these concepts to develop the window join algorithm. We briefly review the material from [4].

A common type of query encountered in telecommunication network performance databases involves aggregating sequences of tuples into larger units. We have seen an example of this type of query in in Example 1.

Example 4: Let us recall the query of Example 1. Using the same schema, we are trying to find the number of packets in each connection (i.e., a connection starts with a `StartCall` packet and ends with a `EndCall` packet). This query can be expressed in the extended multi-feature syntax as:

```
select Source, Dest, ts, count(Y.*)
from Pkt
group by Source, Dest, StartCall, ts ; X, Y
such that (X.Source=Source and X.Dest=Dest and X.EndCall=1 and
                X.ts > ts)
              (Y.Source=Source and Y.Dest=Dest and Y.ts ≥ ts and
                Y.ts ≤ min(X.ts))
having StartCall=1
```

A group is defined by a connection's opening packet (which is the only tuple in the group). The grouping variable X is the collection of all later packets which end calls between this source and destination. Therefore, the earliest member of X ends the connection. The grouping variable Y is the set of all tuples from the source to the destination bewteen the start of the connection and the end of the connection.

The number of entries in the mf-structure H of this example may be very large, forcing it to the disk and greatly increasing the cost of computing the query. However, two basic and related observations improve the performance tremendously.

- The optimizer can exploit the fact that `Pkt` is ordered on `ts` and produce the answer in one pass, instead of three.
- An entry in the mf-structure (i.e. a group, a session) "opens" during the scan and later "closes". In Example 4, an entry opens with a StartCall packet and closes with an EndCall packet. The mf-structure H needs to store only the currently open entries (the maximum number of concurrently open connections is much smaller than the total number of open connections). By using this optimization, we can usually avoid out-of-core processing of the mf-structure.

These two optimizations are discussed in length in the following sections.   □

4.1   Pass-Reducibility

Recall that in Section 3.3, one of the evaluation optimization strategies is to compute aggregates of several grouping variables simultaneously. However, these aggregates can be computed together only if the grouping variables are not ordered by the emf-dependence graph. In this section, we identify the circumstances under which aggregates of two grouping variables can be computed in the same pass even when they are ordered by the emd-dependence graph.

Definition 41: *Assume that $Y \rightarrow X$, $Y \rightarrow X_1, \ldots, Y \rightarrow X_k$ are the dependencies of Y and C_Y is the defining condition of Y. We say that a grouping variable*

Y is pass-reducible to X, if for all groups and all database instances it is possible (i.e. there exists an algorithm) to compute in the same pass the aggregates of X involved in C_Y and the aggregates of Y, given that the aggregates of $X_1, \ldots, X_k$ used in C_Y have been already calculated. This is denoted as $Y \Rightarrow X$.   □

Example 5: Grouping variable Y is pass-reducible to X in Example 4, because for each group, all tuples of Y precede the first tuple of X. Therefore, we can keep counting for Y until the first tuple of X is detected.   □

It is apparent that we must have a principled way (e.g. some syntactic criteria) to identify pass-reducibility among grouping variables. We give here a simple syntactic criterion that covers many extended multi-feature queries similar to 4:

Criterion 41: *Assume that a relation* R *is ordered (ascending) on one of its attributes* t, *Q is an extended multi-feature query and* C_X *denotes the defining condition of grouping variable X in the* such that *clause. Further suppose that* t *is one of the grouping attributes of Q and all defining conditions are conjunctions. and* t *in the* such that *clause changes to* min(X_0.t)*, where* X_0 *is the grouping variable denoting the group (e.g.* X.t $\geq$ t *becomes* X.t $\geq$ min(X_0.t)*.) We can do that since* $t = min(X_0.t)$*. Then:*

a *If* $Y \rightarrow X$ *and one of the subexpressions of* C_Y *has the format* Y.t *op* min(X.t)+c*, where op* $\in \{>, \geq, =\}$ *and c a non-negative integer and no other aggregates of X are mentioned in* C_Y*, then* $Y \Rightarrow X$*. X can be* X_0 *(i.e. the group.)*

b *If* $Y \Rightarrow X$ *due to case (a) and there exists Z such that one of the subexpressions of* C_Y *has the format If* $Y \rightarrow X$ *and one of the subexpressions of* C_Y *has the format* Y.t *op* min(X.t)+c*, where op* $\in \{<, \leq\}$ *and c a non-negative integer and no other aggregates of X are mentioned in* C_Y*, then* $Y \Rightarrow X$*. X can not be* X_0*.*

c *If* $Y \rightarrow X$ *and* C_Y *involves X's aggregates other than* min(X.ts)*, then if syntactically can be proven that X's tuples are preceding Y's tuples, then* $Y \Rightarrow X$*.*

Proof: See [4]. ▮

Example 6: Consider Example 4. The grouping variable X is pass-reducible to X_0 because of case (a) of Criterion 41. Y is pass-reducible to X_0 and X due to cases (a) and (b) respectively. Thus, this query can be answered in a single pass through the data set.   □

In addition, evaluation algorithm 31 must be changed (e.g. keep some additional info on mf-structure's entries) to reflect the fact that a grouping variable Y can be computed in the same pass with another grouping variable X. The revised algorithm is presented in [4].

4.2   Active and Completed mf-Structure Entries

There are cases, as in Example 4 where a group can "close-out" before the end of the scan(s). If it is known somehow that tuples further in the current scan will not affect an entry's grouping variable's output that is being calculated in that scan, then that entry is "completed" with respect to that grouping variable. If an entry becomes "completed" with respect to all grouping variables, then it can be removed from the mf-structure, reducing its size. For one-pass queries such as Example 4, if the output is computed and entries removed as soon as possible, it is likely that the mf-structure can fit into memory. For multiple pass queries, "completed" entries can be transferred to disk, allowing the "active" part of the mf-structure to be in memory.

Definition 42: *An entry h of the mf-structure H of an extended multi-feature query Q is said to be* completed *with respect a grouping variable X, if it is known that tuples further in the current or later scans will not affect the output of X with respect to h. If an entry h is completed with respect to all grouping variables, the entry is called* completed. *Otherwise it is called* active.   □

Once again we must have some syntactic criteria to determine when a grouping variable X closes-out. One way is to know that tuples further in the scan do not affect *the output* of the grouping variable X of an entry (Criterion 42 (a)); a different way is to know that tuples further in the scan will not *belong* to grouping variable X of an entry (Criterion 42 (b).)

Criterion 42: *Assume that a relation* R *is ordered (ascending) on one of its attributes* t, *Q is an extended multi-feature query and C_X denotes the defining condition of grouping variable X in the* such that *clause. Further suppose that* t *is one of the grouping attributes of Q and all defining conditions are conjunctions. and* t *in the* such that *clause changes to* min(X_0.t), *where X_0 is the grouping variable denoting the group.*

a *If X's output consists entirely of* min(X.t), *then the first time that C_X is satisfied for an entry h (on pass degree(X)), h becomes completed with respect to X.*

b *If one of the subexpressions of C_X has the format* X.t *op* min(Z.t)+c, *where op $\in \{<, \leq, =\}$, c a non-negative integer, and Z a grouping variable, then the first time that this subexpression is violated for an entry h on pass degree(X), after the first tuple of X has been identified, h becomes completed with respect to X.*

Proof: See [4]. ∎

Example 7: X's output in Example 4 consists only of min(X.ts). When the first tuple of X for an entry h has been found, then we can "close-out" X with respect to h because of Criterion 42(a): the output of X has been found for entry h.

One of the subexpressions of C_Y in Example 4 is `Y.ts` $\leq$ `min(X.ts)`. The first time that this subexpression is violated for an entry h means that we can "close-out" Y with respect to h because of Criterion 42(b): no tuples later in the scan can belong to Y for h, since their `ts` will be greater or equal than the tuple's `ts` that caused the violation.   □

Attribute `t` could be anything on which the relation is ordered. For example, sales of a company may not be ordered on `date`, but may be ordered on `month` (due to the recording system of the company - batch updates.) Queries having `month` in the grouping attributes can exploit Criteria 41 and 42. Note that these Criteria assume ascending order on attribute `t`. In case of descending order these criteria are still applicable by replacing `min(`X_i`.t)` to `max(`X_i`.t)` and using the inversed operators.

5 The Window Join Algorithm

The processing described in Section 4 is can be viewed as a technique for efficiently computing a self-join on a very large relation (e.g. the example query in this section requires a self join when expressed in SQL, in Example 1). For a final pre-requisite to presenting the window join algorithm, we need to discuss how to handle aggregation over multiple fact tables in EMF SQL. In [8], we show that aggregation over multiple fact tables is simple to define and evaluate. As we show in [5], EMF SQL separates the definition of the groups from the definition of the aggregation. If we associate a grouping variable with a table other than the one that defines the groups, evaluating aggregates is matter of scanning over the new table in Algorithm 31.

Lets consider an (highly simplified) example.

Example 8: Suppose that we are developing an internet telephony application. When a user starts a session, they receive a dynamic address. For usage accounting, we would like to count, for each user session, the number of packets sent in that session. The Sessions relation has the following schema:

> `Sessions(Dynam_Source, Cust_ID, ts, Duration)`

Then, we would use the following query:

```
select Cust_ID, ts, count(X)
from Sessions
group by Cust_ID, ts, Dynam_Source, Duration ; X(Pkt)
such that X.source = Dynam_Source and X.ts ≥ ts and
        X.ts ≤ ts + Duration
```

The syntax `X(Pkt)` indicates that grouping variable X ranges over the Pkt table rather than the Sessions table. The query is evaluated by making a pass through the Sessions table to define the groups, then a pass through the Pkt table to compute the aggregate.   □

The *window join* algorithm is the multi-table analogue of EMF SQL query processing with the optimizations of Section 4 applied, and can be applied when the following criteria holds.

Criterion 51: *Suppose that grouping variable Y is defined over a table other that of X_0 (i.e., the groups). Then, we can apply the window join algorithm for processing Y and X_0 together if:*

a *Y is pass-reducible to X_0.*
b *We have a criteria to determine when all aggregates of Y are completed.*

Suppose that X_0 is defined over S and Y is defined over R. If criteria 51 holds, we can scan S and R together, store a limited portion of S in a hash table, and join the scanned tuples of R with the data in the hash table. That is, conditions a and b define for each tuple s in S a window in R on the tuples that can join with s. The algorithm stored the inverse window of S for the current tuples of R.

The algorithm for performing the join is similar to the one described in [4] (which extends Algorithmm 31 by processing multiple grouping variables in a single scan). However, the algorithm must be extended to scan two relations simultaneously. Let X_0 be defined over S and grouping variable Y defined over R:

a Tuples from S are processed by the logic for X_0, while tuples form R are processed by the logic for Y.
b Let s be the currently scanned tuple from S, and r be the currently scanned tuple from R. Let the criteria that shows pass-reducibility be `Y.ts` *op* `min(`X_0`.ts)+c`. Then we require that $s.ts > r.ts + c$. If *op* is $>$, then we can relax this restriction to $s.ts \geq r.ts$.

Example 9: Lets consider Example 8. By clause a of Criteria 41, X is pass-reducible to X_0. By clause b of Criteria 42, we can determine then X is closed-out, allowing us to reclaim the space for reuse. Therefore, the query can be computed using the window join. Since $c = 0$, we only need to require that if a tuple `t` is being processed from `Pkt`, `t.ts` is less than `s.ts`, where `s` is the last processed tuple from `Sessions`.   □

In most cases, this simple algorithm is can perform a one-pass join and aggregation of R and S using in-memory processing only. Furthermore, the algorithm can be extended in several ways:

Extended Windows: Suppose that one of the defining clauses of grouping variable Y is `Y.ts` *op* `min(`X_0`.ts)-c` for $c \geq 0$. Then we can still apply the window join algorithm, as long as $s.ts > r.ts - c$. We note that we can use this technique to extend the definition of pass-reducibility.

Out-of-core Processing: In order to progress in scanning R, we need to scan S at some minimum rate. Due to memory constraints, we might not be able to create mf-structure entries defined by tuples of S. When an out-of-memory condition occurs, we can write the tuples of S to a temporary file. We then make a second pass to complete the join using the deferred tuples of S (of course, other algorithms are possible). In many cases, we can use the constraints on Y to avoid scanning the entire R table again.

Relational Joins: We can make the output of Y a projection of the attributes of s and r for every time the constraint of Y is satisfied at an mf-structure entry. The total output is the join of R and S, and can be used for further processing. Furthermore, the window join is not restricted to EMF SQL, it can be used in a relational algebra. Given a join expression, if the pass-reducibility and completion criteria can be satisfied (with one of the two joined relations taking the place of X_0), then the window join algorithm can be applied.

6   Related Work

The join algorithm that most closely resembles the window join is the diag-join of Helmer, Westmann, and Moerkotte [7]. These authors observe that fact tables in large data warehouses are usually temporally ordered. Therefore, joins tend to be localized. The diag-join maintains an in-memory window on the smaller table, scans larger table, and performs an in-memory hash join. The window on the smaller table moves to be centered on the expected join location of tuples from the larger table. Multiple passes are used to handle joins of tuples that fall outside the window.

The diag-join and the window join are similar because both exploit the ability to define a localized window on the tables where the tuples will join. The main distinction between the window join and the diag-join is that the window join analyzes the query to determine the window sizes on a per-tuple basis, while the diag-join assumes an a-priori window size. In network analysis applications, the window for some tuples is very large compared to most windows (e.g., very long sessions). The diag-join would require very large windows, while the window join efficiently adapts window sizes at execution time. In addition, the diag-join is an equijoin while the window join is more general, and the window join is designed to be used in conjunction with an efficient large scale aggregation algorithm.

The problem of integrating tertiary storage into database management systems has become an active research problem [1]. In [11], Sarawagi and Stonebraker give query optimization techniques for complex SQL queries on tape-resident data. Myllymaki and Livny [10] have investigated disk-to-tape and tape-to-tape join algorithms that do not make use of indices, which they show are prohibitively expensive. Tribeca is a database for processing very large network traffic datasets [13]. Because the processing is sequential, querying from tape is supported, but the query language is highly restrictive.

A running example in this paper involves the analysis of sequences of tuples. Sequence databases [12] have been proposed as a more convenient and more efficient way to make these queries. Our concern is with a broader range of decision support queries.

7   Conclusions

Analyzing network performance data often requires a complex analysis of multiple very large data sets. We have developed a SQL-like language, EMF SQL, that expresses complex aggregation in a simple and succinct way and which has a fast evaluation algorithm. The EMF SQL evaluation algorithm primarily makes use of sequential scans, allowing queries over tape-resident data.

Network analysis often involves aggregating sequences of related tuples. We have observed that EMF SQL is can express these queries well. In addition, we have developed query optimization strategies to ensure an efficient evaluation.

We observed that the optimization strategies the we developed for sequence aggregation can be applied to joining very large tables that are temporally related. The *window join* algorithm uses the query analysis methods that we developed for optimizing sequence aggregation queries to define the windows in which joins might occur.

We implemented the window join to evaluate several network characterization queries (similar to Example 8) that required a join of two tables. One table was approximately 10 Gbytes, the other approximately 100 Gbytes. As the tables are stored in compressed form, the bottleneck in executing the query was accessing the data rather than performing the join. The queries made one pass through the data sets and executed entirely in memory. Previous attempts to execute the query either required excessive amounts of time (e.g. halted after days of execution), failed after running out of memory, or ran on data sets too small to be useful.

References

1. M. Carey, L. Haas and M. Livny. Tapes hold data too; Challenges of tuples on tertiary storage. In *Proc. ACM SIGMOD*, pages 413–418, 1993.
2. D. Chatziantoniou. Ad-Hoc OLAP: Expression and Evaluation. Submitted for publication (Int. Conf. Data Engineering, 1999), August 1998.
3. D. Chatziantoniou. Evaluation of Ad Hoc OLAP: In-Place Computation. In *ACM/IEEE International Conference on Scientific and Statistical Database Management (to appear)*, 1999.
4. D. Chatziantoniou and T. Johnson. Decision Support Queries on a Tape-Resident Data Warehouse. *IEEE Computer (to appear)*.
5. D. Chatziantoniou, T. Johnson and S. Kim. On Modeling and Processing Decision Support Queries. Submitted for publication, 1999.
6. J. Gray and G. Graefe. The five-minute rule ten years later, and other computer storage rules of thumb. *SIGMOD Record*, 26(4):63–68, 1997.

7. S. Helmer, T. Westmann and G. Moerkotte. Diag-join: An opportunistic join algorithm for 1:N relationships. In *Proc. of the 24th VLDB Conf.*, pages 98–109, 1998.
8. T. Johnson and D. Chatziantoniou. Extending Complex Ad Hoc OLAP. Submitted for publication, February 1999.
9. T. Johnson and E. Miller. Performance Measurements of Teriary Storage Devices. In *24th VLDB Conference*, pages 50–61, 1998.
10. J. Myllymaki and M. Livny. Relational joins for data on tertiary storage. In *Proc. Intl. Conf. on Data Engineering*, 1997.
11. M.S.S. Sarawagi. Reordering query execution in tertiary memory databases. In *Proc. 22st Very Large Database Conference*, 1996.
12. P. Seshadri, M. Livny and R. Raghu. The design and implementation of a sequence database system. In *Proceedings of the 22nd VLDB Conference*, 1996.
13. M. Sullivan and A. Heybey. Tribeca: A system for managing large databases of network traffic. Technical report. Bellcore, 1996.

Assessment of Scaleable Database Architectures for CDR Analysis

An Experimental Approach

Wijnand Derks, Sietse Dijkstra, Willem Jonker, and Jeroen Wijnands

KPN Research, PO Box 15000,
9700 CD Groningen, The Netherlands
{w.l.a.derks, s.j.dijkstra, w.jonker, j.e.p.wijnands}@research.kpn.com

Abstract. This paper describes work in progress on the assessment of scaleable database architectures for CDR analysis.[1] First, the scientific and industrial state-of-the-art of scaleable database architectures is presented. This is followed by a detailed description of experiments carried out on two promising architectures: NUMA and NT-clusters. The paper concludes with a discussion of some early analyses of NUMA experiments.

1 Introduction

The generation of so called Call Detail Records (CDRs) by modern telecommunication switches has opened the way to a whole area of new applications. Most importantly CDRs are the basis for almost all operational billing systems today. And recently, with the advance of database technology for very large databases, new applications based on the analysis of huge amounts of CDRs have become within reach. Examples of such applications are traffic management, fraud detection, and campaign management.

Given the large amounts of data (terabytes of CDRs are normal) there is a need for very large and high performance databases. One way to accommodate these requirements is to use specialised high performance database servers such as for example IBM mainframes, Compaq (TANDEM) or TERADATA machines. However, telecommunication companies are hesitating to make large investments in technologies supporting these new applications. What they are looking for is cost-effective and scaleable technology for realising very large databases in an incremental way.

Given that there is a general trend of commodity hardware and software becoming more and more powerful with respect to both processing power and storage capacity the question arises whether this technology is mature enough to support the specific CDR analysis applications of telecommunication companies.

---

[1] The work described here was carried out in the context of EURESCOM project P817 "Database Technologies for Large Scale Databases in Telecommunication".

W. Jonker (Ed.): Databases in Telecommunications, LNCS 1819, pp. 133–143, 2000.

The research described here addresses the above issue: is cost-effective and scaleable commodity technology mature enough to support large scale CDR analysis? In order to answer this question we took an experimental approach. First, we performed a state-of-the-art study. This study resulted in two promising technologies namely NUMA and NT clusters. Second, we developed an extensive test plan based on requirements from the operator of our mobile network. And finally, we performed two series of extensive experiments on both technologies using again operational data from our mobile operator. At the time of writing this paper, the experiments are in progress, therefore we will present some early results based on a partial analysis of experiments performed.

2   State-of-the-Art

The general approach to supporting very large databases is that of parallel database server architectures. Parallelism is seen as the way to provide for performance, availability, and scaleability at the same time. Roughly speaking there are three main architectures:

Shared-memory (SM): systems with multiple processors, connected by a high bandwidth interconnect through which shared memory and disks can be accessed. Each CPU has full access to both memory and disks.

Shared-disk (SD): systems consisting of several nodes. Each node has one CPU with private memory. Unlike the SM architecture, the shared-disk architecture has no memory sharing across nodes. Each node communicates via a high speed interconnect to the shared disks.

Shared-nothing (SN): systems consisting of several nodes. Each node consists of one or more CPUs with private memory and private storage device. The nodes are interconnected via a network. This network can be both a high speed interconnect or standard technology. The cost of shared-nothing systems can be potentially cheap, using standard commodity components (e.g. PC's). However, mainframe systems also adopt the shared-nothing architecture. Due to their proprietary design their cost is not as low as the table should suggest (see MPP below). The table below gives an overview of the strengths of the different architectures. When looking at commercially available systems we see that (combinations of) the above architectures are present.

Symmetric Multi Processing (SMP). SMP systems use a shared memory model and all resources are equally accessible. The operating system and the hardware are organised so that each processor is theoretically identical. The main performance constraint of SMP systems is the performance of the interconnect. Applications running on a single processor system, can easily be migrated to SMP systems without adaptations. However, the workload must be suitable to

Table 1. Strengths of system architectures

	Cost	DBMS Complexity	Availability	Scaleability
SM	+	-	-	-
SD	0	0	0	0
SN	-	+	+	+

take advantage of the SMP power. All major hardware vendors have SMP systems, with Unix or NT operating systems, in their portfolio. They distinguish in maximum number and capacity of processors; maximum amount of memory; maximum storage capacity. SMP machines are already very common nowadays.

Massive Parallel Processing (MPP). A Massively Parallel Processing system consists of a large number of processing nodes connected to a very high-speed interconnect. MPP systems are considered as Shared-Nothing, that is, each node has its own private memory, local disk storage and a copy of the operating system and of the database software. Data are spread across the disks connected to the nodes. MPP systems are very well suited for supporting VLDBs but they are very expensive because of the need of special versions of the operating system, database software and compilers, as well as a fundamentally different approach to software design. For this reason, only a small top-end of the market uses these systems. Only few vendors have MPP systems in their portfolio. Among them IBM, with its RS/6000 Scaleable POWER parallel system, and Compaq (Tandem) with its Himalaya systems.

Non-Uniform Memory Architecture (NUMA). A NUMA system consists of multiple SMP processing nodes that share a common global memory, in contrast to the MPP model where each node only has direct access to the private memory attached to the node. The non-uniformity in NUMA describes the way that memory is accessed. Somehow, the NUMA architecture is a hybrid resulting from SMP and MPP technologies. As it uses a single shared memory and a single instance of the operating system, the same applications as in an SMP machine run without modifications.[2] The latter advantage makes NUMA a significant competitor to pure MPP machines. Several vendors have computers based on NUMA in their portfolio among which Data General, IBM, ICL, NCR, Pyramid, Sequent and Silicon Graphics.

Clusters. The scaleability limits of SMP systems combined with the need for increases resilience led to the development of clustered systems. A cluster combines two or more separate computers, usually called nodes, into a single system. Each node can be an uni-processor system or an SMP system, has its own main

---

[2] Small modifications will be made to the OS and DBMS, to lower increase remote memory access.

memory and local peripherals and runs its own copy of the operating system. For this reason the cluster architecture is also called a Shared Nothing architecture. The nodes are connected by a relatively high speed interconnect. Commercially available cluster solutions distinguish in the maximum number of nodes and the type of interconnect. In the world of Unix, several cluster solutions are already available (e.g. the SUN Enterprise Cluster). In the World of Windows NT, clusters are yet in their infancy. Whether a cluster behaves like a shared nothing architecture depends on the DBMS architecture. For example, Oracle Parallel Server implements a shared disk architecture on a cluster, whereas DB2 UDB EEE exploits the shared-nothing architecture.

As far as hardware platform support for Very Large Data Bases is concerned, we see the following situation. For very large operational databases, i.e. databases that require heavy updating, mainframe technology (mostly MPP architectures) is by far the most dominant technology. For data warehouses , i.e. databases that provide decision support capabilities, we see already a strong position for the high-end UNIX SMP architectures. The big question with respect to the future is about the role of Windows NT and Intel. Currently there is no role in very large databases for these technologies, however this may change in the coming years. We see Intel based NUMA machines being introduced. These machines are still based on Unix, but suppliers are working on NT implementations. At the same time NT based Intel clusters are at the point of entering the market. As far as database support is concerned, NUMA is more mature and supports major databases like Oracle, DB2, and Informix. For NT clusters, there is a product version available for DB2 UDB EEE from IBM, as well as announcements from Informix (IDS/AD/XP) and Compaq (NonStop SQL).

3   Experiments

Based on the state of the art we decided to carry out two series of experiments on two different architectures: NUMA and NT cluster. The experiments are based on a data warehousing application from the mobile operator of KPN. The experiments consist of a mix of basic and real-life business queries that will test the scalability of the two architectures. At the same time robustness and the ease of management of the system will be considered. Robustness and management issues are among others important to reduce maintenance costs.

3.1   The Database Description

The database contains individual customer information. This information includes data like name, address but also individual CDRs. The amount of CDRs in the database is far more than the amount of customer data. The customer data take 10%, and the CDRs take 90% of the storage. A simplified version of the actual schema is given below.

The experiments are based on actual data, except for the CDRs. The CDRs are generated artificially by a CDR generating program for privacy reasons.

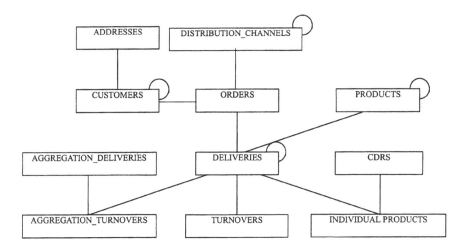

Fig. 1. Database schema used for the experiments

3.2   Description of the Experiments

Three subjects will be investigated:

- *Manageability.* the ease of which the total system is configured and changed. This addresses configuration, loading, backup and change management, and is determined only qualitatively.
- *Scalability.* central are speed-up and scale-up. The speed-up factor of the system will be experimented by scaling the cluster (e.g. number of nodes). The scale-up factor will be experimented by scaling both the database size and the cluster.
- *Robustness.* how the systems can handle failures. On the DBMS level this includes recovery from failures of fragments; on the OS level this includes recovery from operating system shutdown and on hardware level this includes fail-over recovery from node. Robustness will be investigated only qualitatively

The subjects are investigated by performing various queries and installation work on the system. Specifically, nine hardware configurations will be applied to perform concurrency experiments, index creation and marketing queries. In addition the database size is scaled from 40 GB to 80 GB to 160 GB. As the database is mirrored, the total disk space used is 80, 160, and 320 GB.
The experiments are quite diverse and cover all of the three subjects described above. A number of experiments focuss on manageability, inspired by the typical growth path and usage pattern of a CDR database. These experiments address ease of installation of an initial configuration, ease of hardware reconfiguration (i.e. adding and removing nodes, adding and removing disks, switching CPUs

on and off etc.), easy of bulk loads, back-up, and recovery, and ease of database expansion and reconfiguration. The remaining experiments mainly address scalability. These experiments consist of queries derived from typical CDR analysis scenario's. The experiments address indexing, multi-way joins, group-by, concurrent execution of heavy analysis queries, and a number of real-life business queries requiring full table scans. We will discuss the group-by and multi-way join in some more detail. The subject of study is scaleability:

Table 2. Scaleability Analysis Method

Method	Parameters
Investigate the effect of different hardware and database configurations on the response time.	Response time
Identify the bottlenecks in the system during execution.	CPU, disks, memory usage
Check whether the query plan changes when the DB size or hardware configuration changes.	Query plan

The aim of the experiments is to investigate the scalability of typical group-by and multi-way join operations during CDR analysis. The SQL statements used are:

Group-by Query

Select   servmobnr, otherprtynr, sum(to_number(calldur))
From   cdr
where   to_char(strtchrgdate,'hh24') between '07' and '20'
group by servmobnr,otherprtynr

This query computes the total duration for each phone number combination during daytime.

Multi-way Join Query

Select   ads.surname, sum(to_number(calldur))
From   deliveries lvg, customers klt,
*   addresses ads, cdr, individual_products ipt*
Where   cdr.servmobnr = ipt.feature and
*   ipt.lvg_id = lvg.id and cdr.strtchrgdate between*
*   lvg.date_begin and lvg.date_einde and*
*   lvg.klt_id = klt.id and ads.klt_id = klt.id and*
*   ads.tas_code = 'visit'*
group by ads.surname

This query computes the total call duration for each subscriber.

During the experiments the following measurements were carried out:

- wall-clock response times
- utilisation, service times and throughputs of CPU, memory, and disks
- service times and utilisation of kernel and communication overhead

The hypotheses are:

- The query optimiser will always maximise the amount of busy query slaves.
- The response time is N log N related to the DB size.
- The response time is linearly proportional with the number of CPU's.
- The available memory is always fully used.
- The NUMA experiments description

For the NUMA experiments we have chosen the NUMA-Q 2000 system of Sequent Computer Systems. The system runs a, system V based, UNIX operating system called DYNIX/ptx. The database management system is Oracle 8.0.5. The system is configured with four quads (4 CPU's (Intel Pentium 200 MHz) and 1 GB memory each) and in total 48 mirrored disks, each 9.1 GB. This results in an overall capacity of 16 CPU's, 4 GB memory and 436.8 GB of disk space. The system is a so-called cache-coherent NUMA system. Caches are connected by a high speed interconnect, the IQ-link, which takes care of cache synchronisation. The disk subsystem is connected through Fibre Channel Switches and Fibre Channel Bridges as shown below. To cover for disk failures, the disks are mirrored.

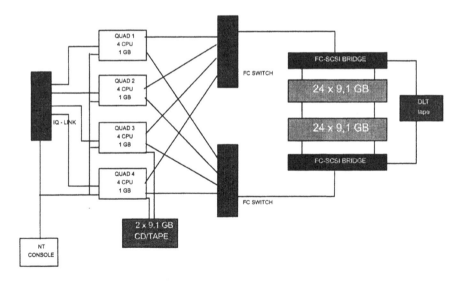

Fig. 2. Architecture of the NUMA system used during the experiments

The NUMA system can be configured in different ways (the inner rectangles represents the memory, the inner squares the CPUs and a white collor indicates "switched off"). We can change the number of quads, the number of CPUs per quad, and the number of disks.

Fig. 3. NUMA configuration for the Quad scaling experiments

Also effects of IQ-link usage and main memory size can be investigated by comparing different configurations. For example running the same query on the three configurations below, allow the investigation of IQ-link.

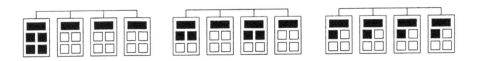

Fig. 4. NUMA configuration for the IQ-link scaling experiments

When we investigate the various options, we find that several are essentially the same. By combining these options we get nine different hardware configurations: five different configurations with four Quads, one configuration with three Quads, two configurations with two Quads and one configuration with one Quad. When we combine the nine hardware configurations with the three database sizes, we end up with 27 configurations in total. In total over 300 individual experiments were performed.

3.3   The NT Cluster Experiments Description

For the NT Cluster experiments we have chosen to use four *IBM Netfinity 7000/M10 systems*. The systems run Windows NT 4.0. The database management system is DB2 UDB EEE. Each machine has four Xeon 450 Mhz CPUs, 2,25 GB memory, 2 4.5 GB disks and 16 9.1 GB disks. This results in an overall capacity of 16 CPUs, 9 GB memory and 618,4 GB of raw disk space. The disks will be configured in a RAID-5 way. A 2.4 Gigabit interconnection of the systems is provided by the *IBM Netfinity SP Switch*. This switch originates from the RS/6000 segment.

4   Early Results and Conclusions

As was stated above at the time of writing this paper only a partial analysis of results has taken place. As a result, no definite conclusions can be drawn yet. To illustrate the specifics of a NUMA configuration we will discuss the analysis of a set of 24 experiments. The experiments used the same business query and were all performed on a 80 GB database. In total 8 different hardware configurations were used and the workload was varied from 1 to 2 to 3 to 4 concurrent executions of the same query. The query used included joins and group by's. Two view definitions were included in the query. The results are shown in figure 5.

The first figure shows lines extrapolated from 12 measured wall clock times on 4 different 4 Quad configurations. The first, Q4P1111, is a configuration of 4 Quads with each 1 processor switched on. The second, Q4P2222, is a configuration of 4 Quads with each 2 processors switched on, etc. The second figure shows lines extrapolated from 12 measured wall clock times on 1, 2, 3, and 4 Quads respectively, where all Quads have all processors switched on.

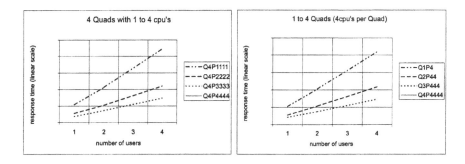

Fig. 5. Scaleability of concurrent users

When looking at scalability, ideally one would like performance to improve linear with system size, i.e. under the same workload two Quads are twice as fast as one Quad. In our experiments we found an improvement factor ranging from 1.91 to 1.94 when comparing Q1P4 to Q2P44, while when comparing Q2P44 to Q4P4444 we found factors ranging from 1.35 (1 user) to 1.69 (4 users). These figures are low. For the single user case, one might argue that the workload is not heavy enough to exploit all four Quads, however for the four users case, a workload of four users of four Quads is comparable to a workload of two users on two Quads. Given that we find a factor 1.91 in that case, the 1.69 has to have a different explanation. Further examination indicates that the disk subsystem might be the bottleneck.

An important aspect of the NUMA architecture is the mechanism for maintaining cache coherency between different Quads. The Sequent system uses a so-called IQ-link for this purpose. We investigated the functioning and role of the IQ-link in our experiments. Comparing the top lines (Q4P1111 and Q1P4) we see a performance advantage for the Q1P4 ranging between 3% and 6%. For the second top lines (Q4P2222 and Q2P44) we see maximum performance difference of 2%. This indicates that for these queries the Oracle / NUMA combination behaves like a real SMP machine.

In order to get a more general feeling for the effect of the IQ-link, we performed a number of experiments on the following configurations Q1P4, Q2P22, and Q4P1111 for various queries. Memory is kept constant over all configurations. Note that, apart from the IQ-link penalty on performance, there is a potential gain due to more I/O bandwidth.

Figure 6 shows the behaviour for the group-by and multi-way join respectively.

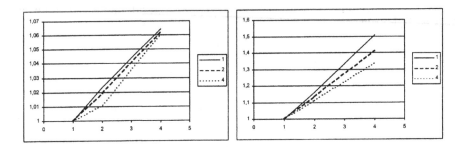

Fig. 6. Scaleability of IQ-link for GB and MJ queries

The query run-times for the Q1P4 configuration have been set to 1. The horizontal axis shows the three configurations Q1P4 at 1, Q2P22 at 2 and Q4P1111 at 4. The different lines represent the 40 Gbyte, 80 Gbyte and 160 Gbyte tests respectively. The values of the Y-axis denote the relative increase in response time for the three hardware configurations. From the first figure one can see that the IQ-link has not much effect on the group-by query, the Q4P1111 configuration is about 6% slower than the Q1P4 configuration. However, for the multi-way join we find the Q4P1111 configuration about 50% slower than the Q1P4 configuration.

The work done so far gives us some early ideas about the applicability of NUMA technology. This needs to be complemented by further analysis and experiments on the NT-cluster technology. Altogether it will give us a good insight in the maturity of the above technology for CDR analysis.[3]

References

1. EURESCOM P817, Deliverable 1, Volume 2, Annex 1 - Architectural and Performance issues (1998)

---

[3] At the time of the workshop the NUMA analysis will be completed, as well as large parts of the execution of the NT-cluster experiments. By that time we will be able to give answers concerning the suitability of the technologies for CDR analysis.

2. Mehta, M., DeWitt, D.J. Data placement in shared-nothing parallel database systems, The VLDB Journal **6** (1), Springer Verlag, Berlin (1997) 53–72
3. Norman, M., Thanisch, P. Parallel database technology: an evaluation and comparison of scalable systems, The Bloor research group, Milton Keynes, UK (1995)
4. Catania, V., Puliafito, A., Riccobene S., Vita, L., Design and Performance Analysis of a Disk Array System, IEEE Transactions on Computers, **44** (10) (1995) 1236-1247
5. Wain,   J.   CC-NUMA   bibliography   (1996)   http://www-frec.bull.fr/OSBU2$_0$/biblio.htm
6. Dolan, D. SMP versus NUMA versus Clustering: Which architecture is best?, Gartnergroup market analysis (1998)
7. Gupta, A. et. al. Performance limitations of CC-NUMA and hierarchical COMA multiprocessors and the flat-COMA solution, Computer Systems Laboratory, Stanford University, Stanford (1992) Technical report CSL-TR-92-524

How to Analyze 1 Billion CDRs per Sec on $200K Hardware

Ian Pattison and Russ Green

TANTAU Software Inc,Max Planck Strasse 36,
Friedrichsdorf,
D-61381 Germany
{Ian.Pattison, Russ.Green}@tantau.com
http://www.tantau.com

Abstract. Modern telecommunication systems generate large amounts of data such as details of network traffic and service usage (call detail records). Amongst other information, these huge databases contain the behavioral patterns of the company's customers. By extracting this data, a telecommunications company (Telco) can better understand the needs of its customers.

Traditional database technology scales to hold vast amounts of data but has severe performance limitations when it comes to analyzing this data. Data mining tools, which often store data in a private representation, offer fast analysis on small data sets but generally do not scale beyond a few million rows.

This paper presents a scalable, parallel data analysis engine capable of processing tens of millions of rows per second per CPU. This technology enables knowledge workers to get sub-second responses to queries that would previously have taken minutes or even hours. . . .

1 Introduction

Modern business and telecommunication systems collect vast amounts of detailed data such as call detail records (CDRs). These call detail records are used for billing purposes and are often stored in order to answer customer queries. This stored data contains a lot of hidden detail about a company's business and the challenge is how to analyze this vast amount of data to discover this valuable information. Modern techniques such as online analytical processing (OLAP) and data mining support the dynamic visualization of data and the finding of hidden patterns in data. However, such techniques normally access a database management system that severely restricts the amount of data that can be processed in a reasonable amount of time due to the performance limitations of such systems.

Most of these database management systems have been designed to efficiently support OLTP applications. There is a major difference between the processing of online transactions and data analysis:

W. Jonker (Ed.): Databases in Telecommunications, LNCS 1819, pp. 144–157, 2000.

- an online transaction usually looks up information that deals with a single unit of work such as detailed customer, service, or call data and may result in an update or insert of a database row. This normally will not affect more than ten or so rows.
- to analyze data, complete tables or large portions of the tables must be scanned. Usually, such analysis involves only a few aspects of the data at a time; for example, relating age and income of customers with their calling patterns. Thus traditional database systems which read in entire rows of data in order to access data from a few columns are very inefficient over large data volumes.

This paper presents a scalable, parallel analysis engine, InfoCharger, whose design goal is to address the needs of analytical processing. InfoCharger allows for efficient storage of data and provides powerful functions that can be used by data mining, OLAP or business intelligence tools.

Section 2 discusses why a traditional database management system is not ideally suited to the requirements of analytical processing. An overview of the architecture of InfoCharger is presented in Section 3 along with details of why this architecture is highly suited to OLAP and data mining. Section 4 describes the storage model used by InfoCharger. Section 5 presents a short case study of a CDR analysis project that used InfoCharger and Section 6 answers the question how to analyze 1 billion rows per second on hardware costing less than $200,000. Approaches other than InfoCharger are considered in section 7.

2   Traditional DBMS and Data Analysis

Traditional database management systems (DBMS) are strong in on-line processing, delivering high performance and transaction protection for multiple concurrent requests. They are optimized to support the complex data structures needed to implement business transactions.

In analytical processing, data structures are usually a lot simpler, with hierarchical structures like star-models or snowflake-models used frequently [4]. Navigation through the models is usually conceptually simple: you need to be able to build aggregates at any level in a hierarchy, and next you need to be able to zoom into interesting parts of the data. The difficulty lies in the vast quantities of data that need to be analyzed: single queries may span millions of rows, and still need to be executed with on-line response times. To make matters more challenging, typical data analysis algorithms generate hundreds of these types of queries. An example of just such an algorithm is an entropy based decision tree [1].

A traditional DBMS uses a row-orientated approach such that all data for a given row is co-located. This is ideal for transactional requests that need to examine many of the columns for a few rows. With a row- orientated storage structure it is not possible to read a partial row. This results in inefficient use of memory and I/O bandwidth when a query accesses only a few columns from each row. Such queries are typical during analytical processing and often access hundreds of millions of rows.

The performance of DBMS systems can be improved by building indices and or pre-calculated summaries. These techniques complicate and slow down data loading but mean that a predefined set of queries can be satisfied efficiently. As an example, the TCP-D benchmark using Oracle version 8 used a total of 715 GB of disk space to store 100 GB of raw data [5].

In many analytical systems, much processing is ad-hoc: there are few pre-defined access paths or pre-defined queries. One needs to be able to analyze data from various points of view with consistent performance. In such an ad-hoc environment it is not possible to pre-build indices or pre-defined aggregate tables: there will always be a need for ones other than those already defined. This section has outlined some of the reasons why a DBMS system with a row orientated design, requiring indices and aggregates is not ideally suited to the demands of analytical processing.

3   InfoCharger - Architecture Overview

InfoCharger is a software product that can store, access and analyze high volumes of data very efficiently. In contrast to traditional database management systems, data structures are optimized for accessing a subset of columns for the millions of rows demanded by analysis and pattern recognition, rather than for accessing individual rows. InfoCharger encodes data automatically as it loads it into its tables and so the required disk space is actually less than the raw data size. This feature not only minimizes disk space and the time to access data on disk, it allows most processing to be done in memory. The performance for typical analytical queries is improved by orders of magnitude compared to traditional database management systems.

The following table compares the performance of InfoCharger with Oracle 8 and SQL Server 6.5. Database consultants Triangle Database Marketing carried out the comparison as part of a report for TANTAU Software Inc [6]. The table shows elapsed times in seconds to perform some of their test queries.

The above test used Triangle's test suite, which uses data to represent a typical data warehouse system of general ledger type data and comprises over 1 million rows.

3.1   System Architecture

InfoCharger has a parallel execution model enabling high performance and scalability. It uses a data parallel approach [3] where rows are distributed over all processors. Each query is sent to all processors, and each processor then applies the same operation to the subset of the rows for which it is responsible. InfoCharger refers to a subset as a partition and each table consists of a set of partitions.

Figure 1 shows an overview of an InfoCharger system.

Table 1. InfoCharger Performance

Query	Grouping attribute	InfoCharger	Oracle 8	SQL Server 6.5
Count(*)	Company	0.05	34.88	5.71
Sum()	Company	0.34	36.38	17.59
Sum()	Glacc	0.37	41.32	17.32
Sum()	Period	0.37	37.23	15.32
Sum()	Type	0.36	37.51	18.28
Sum()	Sector	0.38	36.71	16.46
Sum()	Branch	0.36	39.69	40.42

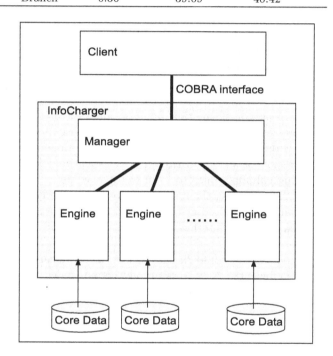

Fig. 1. InfoCharger Architecture

The system consists of a manager and a number of engines. The manager is responsible for accepting requests for clients, it then requests the engines to perform the operation or operations required to satisfy the client request. Each engine is responsible for a single partition and hence an operation will involve as many engines as there are partitions in the table. The engines, in parallel, execute the operation and provide any results to the manager. The manager consolidates the results from the engines and presents these to the requesting client.

InfoCharger supports SMP, cluster and MPP architectures.

3.2   Operational Interface

One of InfoCharger's goals is to enable existing analytical tools to scale to very large data sets. Tools access InfoCharger via a CORBA compliant interface. This interface allows the clients to request aggregates via a general query interface. Histogram and 2-D cross tables are examples of such aggregates.

The interface is extensive and provides methods for use in association analysis and sequence discovery but these are not discussed in this paper.

3.3   Data Preparation and Transformation

Data from an operational system may not immediately be suitable for a data warehouse. It may need cleansing, and more importantly, the data must be translated into a format that has an immediate business meaning.

Pre-processing is required to normalize the data, and import modules typically take care of such normalization. However, over time there may be a need to add new columns, change the meaning of data and modify existing data.

InfoCharger's basic transformation function allows a user to create a new column in a table. InfoCharger stores data column wise (see section 4) and so this requires no database reorganization and can be done quickly. The user specifies an expression of string, arithmetic, statistical and boolean functions to combine existing column values to form the new column. The expression is evaluated at each selected row, the result is converted to the new column's data type if necessary and then stored at that row position in the new column.

InfoCharger has an open interface via which data can be imported into InfoCharger tables.

The storage model used by InfoCharger means that data can be incrementally loaded into tables efficiently. New rows are always appended to existing tables and there are no indices or summary tables to be maintained.

4   InfoCharger - Storage Model

The goal of the storage model is to optimise query performance by:

- holding as much of the data in memory as possible.
- simplifying data structures so that the data can be traversed efficiently.

InfoCharger achieves these goals using column-orientated storage and data compression through encoding. This section explores this storage model and its benefits.

4.1   Column Representation (CORE)

InfoCharger stores the data in a column-orientated fashion rather than the row-orientated fashion used by most DBMS systems.

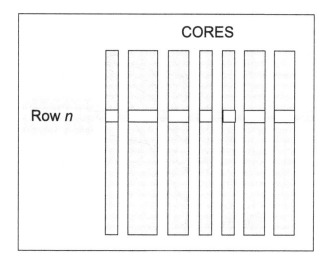

Fig. 2. CORE Representation

Each column is stored separately in its own CORE (Column Represented) Data Object. Figure 2 illustrates this column-orientated approach to storing rows. A CORE can be thought of as a vector where the column value for each row is stored in an element of the vector. The nth row of a table can be re-assembled by selecting the nth element from the set of vectors that represent the table's columns.

Table columns are constructed either as *encoded* COREs or *simple* COREs.

Encoded COREs. Instead of storing the actual values of each row of a column C in a CORE, InfoCharger stores nominal values. These nominal values are offsets into a value table (VAT) for that CORE. The VAT itself stores the unique values encountered in C. In this way, the nominal values in the encoded CORE can be resolved to the actual values by lookup in the VAT.

Figure 3 illustrates the relationship between an encoded CORE and its VAT.

The encoding width of a CORE, i.e. the number of bits in each element of the vector, is as small as the cardinality of the column permits. For example a column with cardinality of 2 where each of the unique values is 100 characters long, will be encoded in a 1-bit wide CORE. The 100 character long actual values will be stored in a VAT.

This type of encoding reduces the memory and disk storage requirements of the source data typically by a factor of between 3 and 7. This contrasts with the extra disk space taken up by indices and summary tables supporting traditional DBMSs as mentioned in section 2. This compression makes processing a lot more efficient as much more information will fit in memory compared to a traditional database management system, thereby enabling more processing to be done without disk accesses.

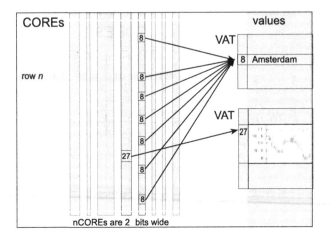

Fig. 3. InfoCharger CORE Encoding

Simple COREs. There are some instances when it is not appropriate to encode a CORE. For example, a column may be of sufficiently high cardinality that encoding may not gain any advantage. In this case, it is possible to store columns as unencoded or simple COREs in which the actual values themselves are held in the COREs. Internally within InfoCharger, the column of a VAT storing the actual values of an encoded CORE is stored as a simple CORE.

Support for Snowflake and Star Schemas. It is often necessary to be able to aggregate data in various dimensions at various levels of granularity. Examples of dimensions are:

- time, where granularity can be hours, days, weeks, months or years.
- geography, where granularity can be switch, region or country.
- customer, where granularity can be individual, family, ZIP-code, city, region or country.

Aggregates need to be built for individual dimensions, like all calls to a country, but also for cross-sections of dimensions, such as calls to a country by week. The database designs that best support these requirements are called the *star-* or *snowflake* model.

To understand how these data models fit into InfoCharger, consider the relationship between an encoded CORE and its VAT. The encoding of a column essentially normalizes the data. All unique values are extracted to a separate table (the VAT) and keys (the nominal values) are stored in the encoded CORE. Actual CORE values can be resolved by equi-joining the CORE to its VAT.

InfoCharger uses a generalisation of this technique to support star- and snowflake schemas. Tables can be declared to be related via a foreign key primary key pair and a relationship type. For example, the customer account column of a CDR table could be related many-to-one to the customer account column of the customer table. In this case, InfoCharger would check that the customer accounts are unique in the customer account table and that all customer accounts in the CDR table exist in the customer account table. The nominal values of the customer account column in the CDR table then define a key to join the CDR table to the customer account table.

InfoCharger's architecture means that this type of table join processing is very fast. Any ad-hoc aggregate can be built at millions of rows per second. In addition, InfoCharger can present a snowflake schema as a single, denormalised table to a front-end tool without the need for explicit schema joins. This is particularly useful for data mining tools that require the data to be presented as a single universal relationship.

4.2   Support for Zooming-In

In many types of analysis, it is necessary to zoom into the most interesting subset of the data, for example, just CDRs for fraudulent calls.

InfoCharger offers efficient zoom-in capabilities by the use of predicates. The caller uses a predicate expression to identify the initial subset of interest. InfoCharger sequentially processes the required COREs and generates a predicate CORE (pCORE) identifying the subset of the rows in the table that meet the requirements of the predicate.

A pCORE defined by a simple predicate expression can be generated at a rate of up to tens of millions of rows per second per CPU.

Aggregate operations can be restricted to a subset of the data by providing a pCORE as an input parameter. The operation to generate a pCORE can itself take a pCORE as an input parameter and hence a subset of interest can be refined without needing to re-apply the initial predicate.

5   InfoCharger - CDR Analysis Example

5.1   Overview

Many Telecom operators (Telcos) already store and process CDRs for purposes other than billing. This is especially true with wireless operators where fraud detection is a vital part of the Telco's infrastructure. The Telcos require fraud detection systems in order to minimize revenue loss through fraudulent calls and to stop legitimate customers being charged for calls they did not make. It is important to avoid the latter as it leads to customer dissatisfaction and hence potentially the loss of customers to competitors.

InfoCharger was used as the analytical engine and CDR store in a proof-of-concept project for a major telecommunications operator.

An overview of the system is given in the following diagram:

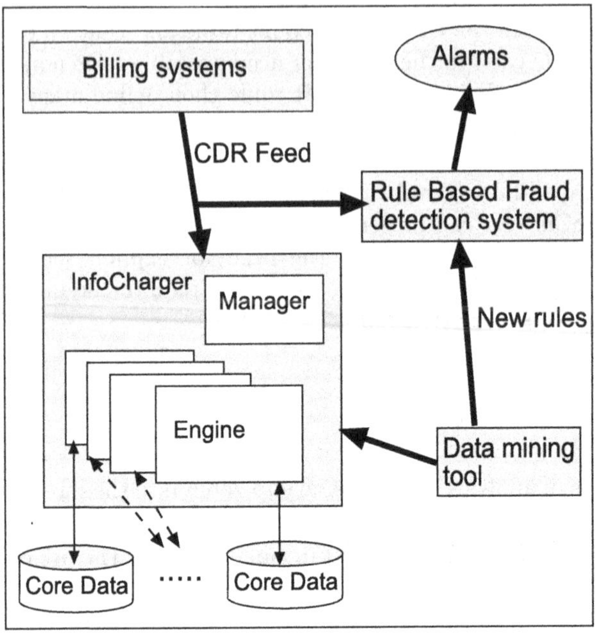

Fig. 4. Overview of Fraud System

The main components of the system were as follows:

- CDR feed. In this case the CDRs were provided from a billing system. However, they could also have been provided directly from the SS7 network if a custom load module had been developed.
- rule based fraud system. A rule based fraud detection system was used to detect suspicious calling behavior and generate alarms. Fraud analysts analyze these alarms to determine whether or not the behavior is actually fraudulent. Although the fraud detection component has a database it does not store all CDRs, it keeps only those CDRs associated with alarms. Specialist fraud analysts defined the *rules* that cause the fraud detection system to generate alarms.
- InfoCharger. All CDRs were stored in InfoCharger.
- data mining tool. A data-mining tool was used by the specialist fraud analysts to search the CDR database for new or refined rules to be used by the fraud detection system. The analysis tool could also be used to analyze historical information not available to the fraud detection system, as the latter does not maintain historical information.

The use of InfoCharger system enabled the fraud analysts to use the data-mining tool against all CDRs in the store rather than just a sample of the CDRs.

In this case the CDRs were for all non-local calls made from a major city for a given period; the total number of CDRs is approximately 45 million. It would have been possible to load many times more CDRs but it was decided that this was not necessary to provide a demonstration of InfoCharger's performance.

5.2 CDR Loading

The fraud detection system requires that the CDRs are provided in a textual representation in normal disk files. For speed of implementation, it was decided to use this same format as the source-input data for InfoCharger thus permitting the use of InfoCharger's built-in text file loader.

Parallel streams reading from separate sets of input files performed the load. The CDRs were loaded into an InfoCharger table with the following attributes (i.e. columns).

- the A number (caller's number).
- the B number (target number).
- the seize time (as an absolute date and time).
- duration.
- area code (of B-number).
- country code (of B-number)
- a set of 24 boolean indicators, e.g. is the call to a premium number, is it an operator call, is it a 3-way call etc.).

The rows were loaded into a table with 4 partitions and a simple hash-based partitioning scheme was used; this partitioning scheme balanced the data evenly over the 4 partitions. The configuration had 4 CPUs, each CPU has one InfoCharger engine and each engine is responsible for a single partition. Queries are executed in parallel with each engine performing the operation on its subset of the data, just over 11 million rows per CPU.

After loading the data, InfoCharger's transformation features were used to transform certain of the attributes into a form better suited for mining. For example, the seize-time was transformed from an absolute to a relative time; the latter is much more interesting when classifying CDRs. The seize-time was transformed into a new attribute - time within day. The flexibility of InfoCharger's transform functionality enabled the time within day to be classified in various different ways, e.g. hour within day, period within day (i.e. morning, evening and night) and also tariff rate period.

6   Performance

The InfoCharger component was installed on a 4 CPU MPP system, a Compaq Himalaya S7004, which uses MIPS R4400 CPUs (125 MHz). Each CPU in the MPP system had 512 Mbytes of memory. The choice of platform was based on the Telco's preferred operating platform and was chosen for its reliability and availability characteristics rather than raw price/performance.

InfoCharger was capable of executing simple queries, such as a histogram, in less than 1.3 seconds. InfoCharger has no built-in aggregates and hence this is a good indication of its raw performance at accessing and processing data. Hence, on the CPUs in this test, InfoCharger is capable of processing approximately 8.5 million rows per second per CPU when generating a histogram.

Another indication of the raw performance of this system would be the time taken to find the first split when building a decision tree. Decision trees are a popular class of data-mining algorithms used by many data mining tools.

The process of finding the first split in a decision tree involves generating a cross-table of each independent variable against the user selected dependent variable. In this case the decision tree involved 29 independent variables. Each 2-D cross table involves accessing two attributes for all 45 million rows: 90 million attribute values must be processed.

Finding the first split in a decision tree involves performing 29 such cross table operations. And therefore involves processing 2,610,000,000 attributes values.

The time taken to find the first split is 75 seconds, which means that each CPU processed an average of 4.35 million rows per second (or 8.7 million attributes per second as two attributes from each row are accessed). Note: this is the total elapsed time and hence includes the time taken to transfer results to the data-mining tool and also includes any processing time spent within the client tool, including screen painting.

In order to build the decision tree at this speed all the COREs for the dependent and independent variables are held in memory. Due to InfoCharger's encoding this is not a significant amount of memory. Each boolean attribute requires 1 bit of memory per row, an attribute with cardinality of 255 or less requires 1 byte per row, cardinality of up to 65,535 requires 2 bytes per column per row and so on.

In this case the decision tree involved 24 boolean attributes, the rest had a cardinality of less than 256. To store all the CORES for 45 million rows in memory therefore required 386 Mbytes (96 Mbytes per CPU).

It is not a requirement to have all these COREs simultaneously in memory. InfoCharger could have performed the same decision tree building with only two COREs simultaneously held in memory, those corresponding to the two attributes in any given 2-D cross table. InfoCharger would then have required to page the correct COREs in and out of memory.

7   One Billion Rows per Second on $200K Hardware

This section builds on the previous section to outline a system that would be capable of processing 1 billion CDRs per second. This system is based on commodity hardware and can built for less than $200,000 of hardware[1].

---

[1] The system outlined is based on hardware prices obtained during October 1998. Today, a similar system could be purchased for less or a faster system could be purchased for the same investment.

The nature of InfoCharger's parallel architecture ensures that near linear scalability can be achieved on most forms of distributed system. The largest configuration tested to date is a 64 CPU system consisting of 16 4-way SMP Windows/NT nodes. Testing on this system demonstrated InfoCharger's near linear scalability:

- adding more CPU power reduces the elapsed time required to execute a query. That is, the data is distributed over more partitions enabling more engines to operate in parallel.
- adding more memory to each CPU increases the number of rows each CPU can efficiently process.

The scalability aspects of InfoCharger are illustrated in the following diagram, which shows the results of tests run on a 16 CPU UNIX system accessing a 10 million row database.

During benchmarking it has been measured that InfoCharger is capable of performing an aggregate at 60 million rows per second on a 300 MHz Pentium II CPU. (Note that this is much faster than the tests on the S7000: the difference is caused by processor architecture, clock speed and compiler technology)

Using InfoCharger's scalability and performance it is possible to build a system capable of analysing 1 billion rows per second, this can be achieved using widely available technology.

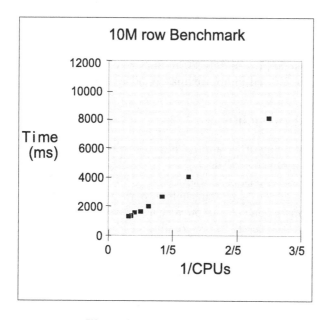

Fig. 5. Scalability Benchmark

Such a system might consist of:

- four 4-way SMP systems giving a total of 16 CPUs.
- each CPU is a 400 MHz Pentium II CPU capable of processing more than 60 million rows per second for a simple aggregate.
- each system has 4 Gbyte of memory providing 1 Gbyte per CPU.
- 5 SCSI disks per system, one partition per CPU and 1 for system disk etc.

Such nodes cost between $30K and $45K each and hence the above system can be built for under $200K.

The near linear scalability properties of InfoCharger ensure that the above system would be capable of performing simple aggregates at more than 1 billion rows per second. Using InfoCharger's measured results for 300Mhz processors as a baseline, a system the same as above but using the such slower 300 MHz processors would be capable of processing 960,000,000 rows per second for a histogram operation. As the above system is priced and configured for the faster 400 MHz processors, InfoCharger would be capable of processing at least 1 billion rows per second given the assumption that at least a 5performance improvement will be measured using the 400 MHz processors.

In the previous section building a decision tree from some CDR information was discussed. This decision tree involved 29 independent variables and the dependent variable.

With 1 billion CDRs rather than 45 million, holding all these COREs in memory would require approximately 536 Mbytes of memory per CPU which is well within the configured 1 Gbyte of memory per CPU. Performing a histogram on an attribute of these one billion CDRs would take approximately 1 second.

Finding the first split would involve 29 2-D cross tables and access a total of 29,000,000,000 rows (58,000,000,000 attribute values). In this configuration each CPU is responsible for 1/16 of the rows. Each CPU is capable of performing a 2-D histogram at approximately 30 million rows a second and hence the first split could be discovered in approximately 60 seconds.

8   Other Approaches

There have been many other approaches to solving the problems posed by large scale, high performance analysis.

In the data warehousing market, most relational DBMS vendors use built aggregates and summaries to achieve high performance. This technique requires careful process management and on very large systems can prove to be too difficult to manage or too costly to pre-compute. Cube data models (NCR Teradata) work well in these environments but don't have anywhere near the performance of in-memory systems.

A number of in-memory databases have been built to attack the performance problem. Timesten (http://www.timesten.com/press/pr990315c.html) show how they can achieve over 1 million operations per minute on a 4 CPU Windows/NT system. These results show how ordinary DBMS systems can improve their performance dramatically if good use is made of memory.

InfoCharger is able to improve by another two orders of magnitude by optimizing for the sorts of queries required for typical data analysis. An unaudited (to be submitted) InfoCharger run against the Drill Down Benchmark [2] showed a factor of three improvement against another specialist query in memory database.

9 Summary

Traditional database systems are designed to satisfy the requirements of on-line transactional processing (OLTP) systems. Although such DBMS systems have been enhanced to improve their performance when processing analytical queries their performance is still inadequate when accessing tens or hundreds of millions of rows.

InfoCharger provides an alternative design for a database engine whose design goal is to satisfy the requirements of analytical tools. As such, it can aggregate rows at 60 million rows per second per CPU. InfoCharger's architecture provides near linear scalability and can be installed on SMP, cluster and MPP architectures.

The combination of InfoCharger's performance and scalability with today's hardware prices mean that it is possible to build a system that is capable of aggregating CDRs at a rate of more than 1 billion per second.

InfoCharger's real strength lies in its ability to show superb performance and scale to very large systems. Its high performance load and data preparation and the ability to present schema in the form needed by a front-end tool combine to make InfoCharger an excellent data analysis engine.

References

1. Berry, J.A. Michael, Linoff, Gordon: Data Mining Techniques. John Wiley & Sons (1997).
2. Boncz,Ruhl,Kwakkel: The Drill Down Benchmark. Proc. of 24th VLDB Conf., 1998.
3. Chattratichat, J. et al: Large Scale Data Mining: Challenges and Responses. Proceedings KDD '97.
4. Inmon, W.H.: Building the Data Warehouse. John Wiley & Sons, (March 1996).
5. TPC-D benchmark for ProLiant 65000 6/200-1, report data 19-september-1997.
6. Triangle Database Marketing: Report produced for internal use of TANTAU Software Inc. June 03, 1999.

A Distributed Real-Time Main-Memory Database for Telecommunication

Jan Lindström, Tiina Niklander, Pasi Porkka, and Kimmo Raatikainen

University of Helsinki, Department of Computer Science
P.O. Box 26 (Teollisuuskatu 23), FIN-00014 University of Helsinki,Finland
{jan.lindstrom,tiina.niklander,pasi.porkka,kimmo.raatikainen}@cs.Helsin-
ki.FI

Abstract. Recent developments in network and switching technologies
have increased the data intensity of telecommunications systems and
services. The challenges facing telecom data management are now at
a point that the database research community can and should become
deeply involved.

In the research project RODAIN we have developed a database architecture that is *real-time, object-oriented, fault-tolerant,* and *distributed.*
It implements object model of ODMG with real-time extensions of our
own. Our prototype of RODAIN database architecture supports real-time
scheduling of real-time and non-realtime transactions based on transaction's deadline and criticality value. The RODAIN concurrency control
is based on an optimistic method which is extended with relaxed serializability and semantic conflict resolution.

1   Introduction

Recent developments in network and switching technologies have increased the
data intensity of telecommunications systems and services. This is clearly seen
in many areas of telecommunications including network management, service
management, and service provisioning. For example, in the area of network management the complexity of modern networks leads to a large amount of data
on network topology, configuration, equipment settings, and so on. In the area
of service management there are customer subscriptions, the registration of customers, and service usage (e.g. call detail records) that lead to large databases.

The integration of network control, management, and administration also
leads to a situation where database technology becomes an integral part of the
core network; for example in architectures like TMN [15], IN [8], and TINA [3].
The combination of vast amounts of data, real-time constraints, and the necessity
of high availability creates challenges for many aspects of database technology
including distributed databases, database transaction processing, storage and
query optimization.

Until now, the database research community has only paid a little attention
to data management in telecommunications and has contributed little beyond
core database technology. The challenges facing telecom data management are

W. Jonker (Ed.): Databases in Telecommunications, LNCS 1819, pp. 158–173, 2000.

now at a point where the database research community can and should become deeply involved.

The performance, reliability, and availability requirements of data access operations are demanding. Thousands of retrievals must be executed in a second and the allowed down time is only a few seconds in a year.

Telecommunication requirements and real-time database concepts are somewhat studied in the literature [5, 2, 22, 4]. Kim and Son [25, 21] have presented a *StarBase* real-time database architecture. This architecture has been developed over a real-time micro kernel operating system and it is based on a relational model. Wolfe & al. [44] have implemented a prototype of object-oriented real-time database architecture *RTSORAC*. Their architecture is based on open OODB architecture with real-time extensions. The database is implemented over a thread-based POSIX-compliant operating system. Another object-oriented architecture is presented by Cha & al. [6]. Their M^2RTSS-architecture is a main-memory database system. It provides classes, that implement the core functionality of storage manager, real-time transaction scheduling, and recovery.

Dali [20] is a main memory storage manager, that can be uses as a memory manager in any single main memory database system. *BeeHive* [39] is a real-time database supporting transaction and data deadlines, parallel and real-time recovery, fault tolerance, real-time performance and security. *ClustRa* [13, 43] is a database engine developed for telephony applications like mobility management and intelligent networks.

The RODAIN[1] database architecture is a *real-time, object-oriented, fault-tolerant*, and *distributed* database management system. RODAIN is designed to fulfill the requirements of a modern telecommunications database system. The requirements are derived from the most important telecommunications standards including Intelligent Network(IN), Telecommunications Management Network (TMN), and Telecommunication Information Networking Architecture(TINA). The requirements originate in the following areas: real-time access to data, fault tolerance, distribution, object orientation, efficiency, flexibility, multiple interfaces, and compatibility [33, 41].

In RODAIN we have developed a real-time main-memory database architecture that is an object-oriented database using an object model of ODMG with real-time extensions of our own. The additional characteristics are designed so that the original ODMG model is a true subset of our extended model. The essence of the extensions is to guarantee that the real-time scheduler and the concurrency controller have enough knowledge to overcome the problems due to heterogeneous lengths of transactions. The prototype implementation of the RODAIN database architecture supports real-time scheduling of real-time and non-real-time transactions.

---

[1] RODAIN is the acronym of the project name *Real-Time Object-Oriented Database Architecture for Intelligent Networks* funded by Nokia Telecommunications, Solid Information Technology Ltd., and the Finnish Technology Development Center (TE-KES).

In telecommunication applications, database transactions have several different characteristics. In RODAIN we have studied two basic types of transactions: service provision transactions (IN) and service management transactions (TMN). The IN transactions are expressed as firm real-time transactions whereas the TMN transactions are expressed as non-real-time transactions. Traditional concurrency control methods use serializability as the correctness criterion when transactions are concurrently executed. However, strict serializability as the correctness criterion is not always suitable in real-time databases, in which correctness requirements may vary from one type of transactions to another and from one type of data to another. Some data may have temporal behavior, some data can be read although it is already updated but not yet committed, and some data must be guarded by strict serializability. These conflicting requirements must be solved through using a special purpose concurrency control scheme. Therefore, the RODAIN concurrency control is based on an optimistic method which is extended with relaxed serializability and a semantic conflict resolution.

An open database management system must support several interfaces to the database. Therefore we have studied interfaces for CORBA, Intelligent Network Application Protocol (INAP), Data Access Protocol (DAP), and Common Management Information Protocol (CMIP).

Any database used in many telecommunication services must also be continuously available. The "official" ITU requirements allow the down time to be only a few seconds per failure. The high availability of the RODAIN database system is achieved by using two separate nodes with their own copies of the full database. One node, called the primary node, executes the database updates. The other node, called the mirror node, monitors the changes in the database content and is ready to take the update responsibilities if the primary node fails.

The rest of the paper is organized as follows. In 2 we briefly summarize the key requirements for a database that is suitable for telecommunication applications. The architecture of the RODAIN Database Management System is presented in 3. In 4 we have collected the needed application interfaces. Chapter 5 covers the transactions and their processing. A more detailed description of the concurrency control is given in 6.

2   Database Requirements in Telecommunications

Below, we summarize the key requirements including *data distribution, data replication, object orientation, object directories, multiple application interfaces, fault tolerance,* and *real-time transactions.* A more detailed description can be found in [42].

Data Distribution. A set of database nodes may cooperate to accomplish database tasks. A node may request information from other nodes in the case information needed is not locally available. It is possible to implement the underlying database system without data distribution. In such an implementation all applications use a single database server. This differentiates logical data distribution from physical data distribution among database nodes. The former is necessary but the latter is a decision internal to the design of the database architecture. Our current belief is that only a few requests need to access more than one database node.

Data Replication. It is stated in ITU-T Recommendation Q.1204 [18] that a database node contains customer and network data for real-time access. We believe that currently most distributed operations are reads. Therefore, replication is an effective way to speed up distributed operations.

Object Orientation. It is a common belief that the best long term telecommunications architectures are object oriented. The implication is that a database system must support object access. In other words, the database system must either be object oriented or have an object interface.

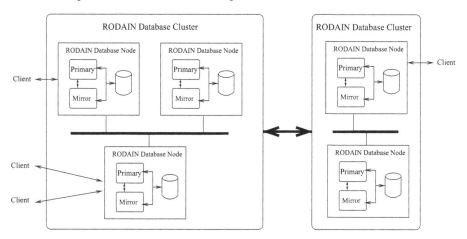

Fig. 1. RODAIN Database Clusters.

Object Directories. The conceptual model of IN Service Control Function (ITU-T Recommendation Q.1214 [16]) defines a Service Data Object Directory that is used to access data stored in the Service Data Functions. This implies that object directories must be supported. The invocation mechanism of CORBA is a good alternative to implement the function.

Multiple Application Interfaces. In telecommunications different architectures define different access interfaces. The IN Capability Set 1 defines IN Application Protocol (INAP). TMN has its own access methods based on the OSI (X.700) management protocols. The current definition in TINA is based on TINA ODL which is quite similar to the OMG IDL interface. Therefore, the OMG IDL interface is also needed. In addition, theOQL interface—probably without response-time guarantees—would be convenient for ad hoc queries and database maintenance.

Fault Tolerance. Real-time access implies that data must be continuously available. The result of the implications is that the database system must be fault tolerant. In the current definitions the maximum allowed down time is a few seconds a year.

Real-time Transactions. Although the real-time data access as stated in ITU-T Recommendation Q.1204 does not directly imply that the underlying database system must support real-time transactions, we believe that the most convenient way to support real-time access to data is to use a real-time database system. In telecommunications we will need both soft and firm transactions. We do not believe that hard transactions will be used in the near future because systems supporting hard transactions are too expensive for open telecommunication markets.

3   Overview of the RODAIN Database Architecture

The RODAIN Database Architecture is a hierarchical distributed database architecture. The RODAIN Database Nodes are linked together to form database clusters. Furthermore, the database clusters are then connected together to allow access between different, possibly heterogeneous, distributed databases (Figure 1).

RODAIN Database Nodes within one database cluster share common metadata. They also have access to a global dictionary in order to locate any data item. Different telecommunication applications as clients of RODAIN Database Cluster can access any one of the Database Nodes. Each node serving a request can fully hide data distribution by providing access to all data items within the cluster. For most time-critical transactions the RODAIN Database offers the possibility to learn the fastest access point of each time-critical data item.

RODAIN Database Clusters do not necessarily share common metadata. Instead, schema translations may be needed when data items from remote cluster are accessed. It should also be noted that no assumption of real-time behavior of the remote clusters can be made since the remote cluster usually belongs to a different administration domain. Therefore, we assume that the communication between clusters will be based on standard communication protocols and object access models like the ones used in CORBA [30]. In this way a remote cluster does not have to be a RODAIN Database Cluster. In fact a remote cluster can be any object or relational database.

The data in each Rodain Database Node is divided into two parts; each data item belongs to hot data or to cold data [35]. The data items are stored in different databases in the Rodain Database Node; hot data is stored in a main-memory database and cold data is stored in a disk-based database. All updates in the hot data are done in the main memory. A transaction log of hot data is maintained to keep the database in a consistent state. A secondary copy of hot data is located in a mirror node and only a backup copy is maintained on the disk. Since cold data is stored in a disk-based database we use a hybrid data management method that combines a main-memory database and a disk-based database.

RODAIN Database Nodes that form a RODAIN Database Cluster are real-time, highly-available, main-memory database servers. They support concurrently running real-time transactions using an optimistic concurrency control

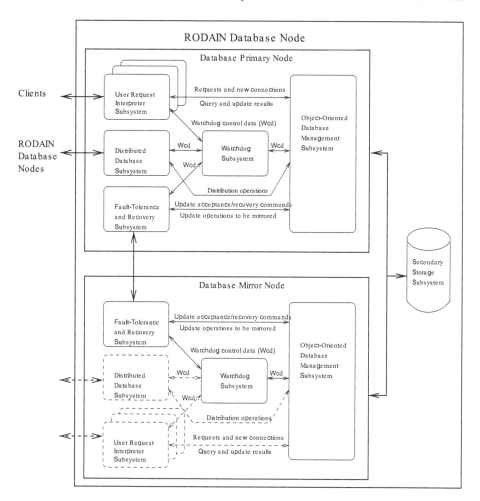

Fig. 2. Rodain Database Node.

protocol with deferred write policy. They can also execute non-real-time transactions at the same time on the database. Real-time transactions are scheduled based on their type, priority, mission criticality, or time criticality.

In order to increase the availability of the database each Rodain Database Node consists of two identical co-operative nodes. One of the nodes acts as the Database Primary Node and the other one, Database Mirror Node, is mirroring the Primary Node. Whenever necessary, that is when a failure occurs, the Primary and the Mirror Node can switch their roles. When there is only one node in function we call it a Transient Node. A Transient Node behaves like a Primary Node, but it is not accompanied by a Mirror Node and, therefore, it behaves like a stand-alone database system. The role of Transient Node is designed to be temporary and used only during the failure period of the other node.

The Database Primary Node and Mirror Node use a reliable shared Secondary Storage Subsystem (SSS) for permanently storing the cold data database, copies of the hot data database, and log information. The nodes themselves are further divided into a set of subsystems (Figure 2) that perform the needed function on both nodes. Below, we will shortly summarize the function of each subsystem.

User Request Interpreter Subsystem. The Rodain Database Node can have multiple application interfaces. Each interface is handled by one specific User Request Interpreter Subsystem. It translates its own interface language into a common connection language that the database management subsystem understands. The URISs on the Primary Node are active and communicate with the clients. On the Mirror Node the URISs are not needed.

Distributed Database Subsystem. A Rodain Database Node may either be used as a stand-alone system or in co-operation with the other autonomous Rodain Database Nodes within one RODAIN Database Cluster. The database co-operation management in the Database Primary Node is left to the Distributed Database Subsystem (DDS). The Distributed Database Subsystem on the Mirror Node is passive or non-existent. It is activated when the Mirror Node becomes a new Primary or Transient Node.

Fault-Tolerance and Recovery Subsystem. The FTRS on both nodes controls communication between the Database Primary Node and the Database Mirror Node. It also co-operates with the local Watchdog Subsystem to support fault tolerance.

The FTRS on the Primary Node handles transaction logs and failure information. It sends transaction logs to the Mirror Node. It also monitors the Mirror Node. When it notices a failure it reports that to the Watchdog Subsystem for changing the role of the node to a Transient Node. On the Transient Node FTRS stores the logs directly to the disk on the SSS.

The FTRS on the Mirror Node receives the logs sent by its counterpart on the Primary Node. It then saves the logs to disk on Secondary Storage Subsystem and gives needed update instructions to the local Database Management Subsystem. When it notices that the Primary Node has failed, it informs the local Watchdog Subsystem to start the role change.

Watchdog Subsystem. The Watchdog subsystem watches over the other local running subsystems both on the Primary and on the Mirror Node. It implements a subset of watchd [12] service. Upon a failure it recovers the node or the failed subsystem. Most subsystem failures need to be handled like the failure of the whole node. The failure of the whole node requires compensating operations on the other node. On Primary Node when the failure of the Mirror Node is noticed, the WS controls the node change to the Transient Node. This change affects mostly the FTRS, which must start storing the logs to the disk. On the Mirror Node the failure of the Primary Node generates more work. In order to change the Mirror Node to the Transient Node the WS must activate passive subsystems such as URIS and DDS. The FTRS must change its functionality from receiving logs to saving them to the disk on SSS.

Object-Oriented Database Management Subsystem. The OO-DBMS is the main subsystem on both Primary Node and Mirror Node. It maintains both hot and cold databases. It maintains real-time constraints of transactions, database integrity, and concurrency control. It consists of a set of database processes, that use database services to resolve requests from other subsystems, and a set of manager services that implement database functionality. The Object-Oriented Database Management Subsystem needs the Distributed Database Subsystem, when it can not solve an object request on the local database.

4   Usage Interfaces

As described in Section 3, there are multiple URISs in the RODAIN architecture. For each protocol allowed to access RODAIN database system, there must be a URIS. Each URIS receives queries and service requests presented in one specific protocol and transforms these queries and requests into a language that the DBMS understands.

An open database management system must support several interfaces to the database. We have studied requirements set by two Intelligent Network's protocols. *Intelligent Network Application Protocol (INAP)* is described in the capability set 1 (CS-1) [17]. It introduces a *Service Data Function (SDF)*, which provides the function needed for storing, managing and accessing information in the network. CS-2 introduces *Data Access Protocol (DAP)* [19]. IN CS-2 defines the rest of the IN services. The enhancements of the DAP include authentication, security, assignment of access rights, the control user's data access and block data access.

An interface for OSI's *Common Management Information Protocol (CMIP)* [14] is also studied. With the CMI protocol, the user can insert an information tree into RODAIN and manage it. With CMIP the user can successfully make connections to the database and manage the information tree, and manipulate the information in it.

CORBA [28, 30] offers means for distribution and builtin connections to several different platforms. It has also gained growing interest in telecommunication. Therefore, we have evaluated the integration of CORBA and DBMS. As a result, we have specified the requirements for an *Object Database Adapter (ODA)* [31]. The direct invocation of database objects from a CORBA application is an important performance issue. It would be impossible for the CORBA to work properly if all database objects were to be registered as CORBA objects. It would cause severe overhead for the CORBA system. An ODA should provide both static and dynamic invocation (DII) of database objects. This is achieved by introducing the interfaces of database objects into CORBA. Plenty of the functionality provided by CORBA is defined in different service specifications [30]. Some of the services include functionality that ODA should also provide. In Figure 3 the basic architecture of the RODAIN Object-Oriented Database Adapter (ROODA) is shown. The names Persistent Object Service (*POS*), Object Transaction Service (*OTS*) and Object Query Service (*OQS*) refer to the corresponding services. *DII* and *DSI* refer to dynamic invocation and skeleton interfaces. It should be noted that, in the context of ROODA only certain parts of these services are implemented.

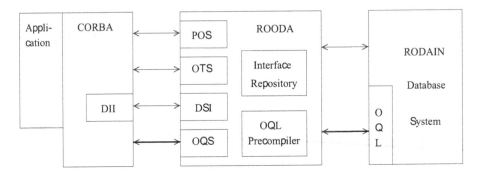

Fig. 3. CORBA, OODA and RODAIN interaction.

The Persistent Object Service (POS) provides a common interface to the mechanism used for retaining and managing the persistent state of objects. The client who has requested or created an object, may make it persistent by invoking objects introduced in the POS interface. For this service, ROODA acts as a *persistent data service (PDS)* and the RODAIN DBMS as a *datastore*. However, in the version 2.2 of CORBA, the POS is abandoned and a *Persistent Object Adapter (POA)* is taking care of persistence [29].

The Object Transaction Service (OTS) provides transaction synchronization across the elements of a distributed client/server application. A transaction can involve multiple objects performing multiple requests. OTS places no constraints on the number of objects involved, the topology of the application or the way in which the application is distributed across the network. However, there is one restriction for the participating objects. The objects called within OTS must be either transactional or recoverable for that operation to display transaction semantics [38].

Dynamic Invocation and Skeleton Interfaces (DII & DSI) provide means for dynamic object invocation. DII allows clients to access objects, which were not available in the system when the client was compiled. DSI is the server's part of the dynamic invocation procedure. They are required in ROODA, since the database evolves constantly.

Object Query Service (OQS) [30] provides predicate-based queries of selection, insertion, updating, and deletion on collections of objects. Operations are executed to *source collections* and they may return *result collections* of objects. The result collections may be either selected from the source collections or produced by query evaluators. Queries may also be set on the result collections.

The OQS consists of two different interfaces; the query interface and the interface for iterable collections. CORBA objects may participate in the OQS in two different ways. All CORBA *objects as themselves* are queryable. A query may be set on an object's attributes, relationships, or methods. CORBA objects can also be *members of collections*. With the iterable collection interface queries may be set efficiently on these collections.

OQS may also set queries to objects that are not managed by CORBA. Usually these external objects reside in a database. CORBA can access database objects basically only one at a time, since CORBA handles single objects. If the database management provides a class collection, objects of that class may be requested by CORBA. Efficient querying of objects from an external database requires the use of the database's native query language. Hence, we must provide an efficient OQL interface in the RODAIN DBMS.

5   Transaction Processing

There are three problems that any real-time database system must deal with: 1) resolving resource contention, 2) resolving data contention, and 3) enforcing timing constraints. In the RODAIN database transactions to be executed are classified according to their deadline and to their importance. Real-time transactions differ from traditional transactions in many ways. In traditional database systems the main goal of transaction scheduling is to maximize the throughput of transactions. In real-time database systems the goal is to maximize the number of transactions that are completed before their deadlines. Predictability of transaction lengths is one of the most important prerequisites for meeting the deadlines in real-time systems.

In real-time database systems transactions must be controlled by a real-time scheduler. A scheduling priority for each transaction is automatically evaluated and modified during the run time. The evaluation process is based on the characteristics of the transaction, the validity of the accessed data object, and the dynamic behavior of the system.

The goal of real-time scheduling is to maximize the number of transactions that are completed before their deadlines. However, this is not always possible. In these situations transactions having low values of the transaction_importance property are sacrificed in favor of those having high values. In the RODAIN database, transactions arrive asynchronously. Due to the nature of telecommunication systems the arriving rate of transactions may vary within a very large bounds. Furthermore, the length of each transaction and its data accessing patterns may vary in an unpredictable way. Therefore, there are sometimes situations when a DBMS load temporarily exceeds its processing capacity. In the literature, this is called an overload situation.

In the RODAIN database we use an adaptive overload prevention policy. A principle of this policy is to limit the number of active Transaction Processes. The present limit is based on system load, which is observed in periodic basis. The metric of system load is how many transactions miss their deadline within an observation period. Another factor for limit calculation is how many aborts are made due to concurrency control.

Transaction processing in the RODAIN system is a task which requires the co-operation of numerous processes. These processes are located 1) in the Object-Oriented Database Management Subsystem (OO-DBMS), 2) outside the OO-DBMS but inside the RODAIN node, and 3) outside the RODAIN node. Transactions can directly modify local and remote data, call (invoke) local and remote methods, and commit or abort modifications done in local node or remote nodes.

During the last 10–15 years several algorithms have been developed to schedule real-time transactions (e.g., [45]). The main goal of the real-time scheduling algorithms is to guarantee that as many transactions as possible will meet their deadline. The scheduling algorithms behave differently depending on both data and processing resource contention [1]. For the RODAIN database system the challenge in real-time scheduling is the fact that existing real-time scheduling algorithms do not guarantee that non-real-time transactions will receive enough resources to complete.

In telecommunication applications, database transactions have several different characteristics [34]. In RODAIN database system two transaction types have been studied: service provision (IN) and service management (TMN) transactions. IN transactions are used to access data of one customer. TMN transactions are used to update database contents widely, for example, when adding new users or configuring user profiles.

IN type transactions are short reads or updates, which usually affect a few database objects. Typically the object is fetched based on the value of its key attribute value. Transaction atomicity and consistency requirements can sometimes be relaxed [41]. However, IN transactions have quite strict deadlines and their arrival rate can be high, but most IN transactions have read-only semantics. In existing IN applications, the ratio of write transactions is approximately 1 to 10 percent [41]. In RODAIN transaction scheduling IN transactions are expressed as firm real-time transactions.

TMN type transactions have opposite characteristics from IN transactions. They are long updates which write many objects. Strict serializability, consistency and atomicity are required for TMN transactions. However, they do not have explicit deadline requirements. Thus, TMN transactions are expressed as non-real-time transactions.

Most database systems offer either real-time or fair scheduling for concurrent transactions. These aspects often conflict with each other. To fulfill telecommunication database requirements conflicting transaction types must be scheduled simultaneously in the same database. Short IN transactions must be completed according to their deadlines as well as long TMN transactions must have enough resources to complete.

The RODAIN scheduling algorithm, called *FN-EDF*, is designed to support simultaneous execution of both firm real-time and non-real-time transactions. In RODAIN firm deadline transactions are scheduled according to the EDF scheduling policy [27, 11]. The FN-EDF algorithm guarantees that non-real-time transactions receive a respecified amount of execution time.

The FN-EDF algorithm periodically samples the execution times of all transactions. The operating system scheduling priority of a non-real-time transaction is adjusted if its fraction of execution time is either above or below the respecified target value. For each class of non-real-time transactions $(c = 1, \ldots, C)$ the target value is given as a system parameter γ_c, which can be changed while the system is running.

The scheduling of transaction processes is based on the FIFO scheduling policy provided by the Chorus operating system [32]. A continuous range of priorities is reserved for transactions. For firm deadline transactions the priority is assigned once and subsequent transactions receive priorities based on previous assignments. When a non-real-time transaction is started, it receives the lowest priority of the priority range. During adjustment phases the priority is raised until the transaction receives the deferred fraction of execution time. Transaction processing of the RODAIN DBMS is discussed in more detail in [23].

6   Concurrency Control

Traditional concurrency control methods use serializability [7] as the correctness criterion when transactions are executed concurrently. However, strict serializability as the correctness criterion is not always the most suitable one in real-time databases, in which correctness requirements may vary from one type of transactions to another. Some data may have temporal behavior, some data can be read although it is already written but not yet committed, and some data must be guarded by strict serializability. These conflicting requirements may be solved by using a special purpose concurrency control scheme.

A heterogeneous transaction behavior also introduces problems to concurrency control. In traditional databases, database correctness is well defined and homogeneous between transactions. However, a demand for strict database correctness is not applicable for real-time databases, where correctness requirements can be heterogeneous. Real-time data is often temporal and neither serializing concurrency control nor full support for failure recovery is required because of the overhead [40]. This has lead to ideas such as *atomic set-wise serializability* [37], *external versus internal consistency* [26], *epsilon serializability* [36], and *similarity* [24]. Graham [9, 10] has argued that none are as obviously correct, nor as obviously implementable, as serializability.

Due to the semantic properties of telecommunications applications, the correctness criterion can be relaxed to so called *semantic based serializability*. We divide the semantic based serializability into two parts. Firstly, we define a temporal serializability criterion called *τ-serializability* [34], which allows old data to be read unless the data is too old. Secondly, we use a semantic conflict resolution model that introduces explicit rules, which are then used to relax serializability of transactions. The first method reduces the number of read-write conflicts whereas the second one reduces the number of write-write conflicts.

We use the τ-serializability as a correctness criterion to reduce read-write conflicts. Suppose that transaction T_A updates the data item x and receives a write lock on x at time t_a. Later transaction T_B wants to read the data item x. Let t_b be the time when T_B requests the read lock on x. In τ-serializability the two locks do not conflict if $t_a + \min(\tau_b, \tau_x) > t_b$. The tolerance $\min(\tau_b, \tau_x)$ specifies how long the old value is useful, which may depend both on data semantics (τ_x) and on application semantics (τ_b).

These novel ideas are utilized with the optimistic concurrency control protocol designed for RODAIN. Optimistic concurrency control protocols have the nice properties of being non-blocking and deadlock-free. These properties make them especially attractive for RTDBS. As the conflict resolution between the transactions is delayed until a transaction is near completion, there will be more information available for making the choice in resolving the conflict. We have presented a method to reduce the number of transaction restarts and a new optimistic concurrency control protocol, called OCC-DATI.

7   Conclusion

We have described the RODAIN database architecture and main parts of its prototype implementation. The RODAIN database architecture is designed to meet the challenge of future telecommunication systems including Intelligent Networks, Telecommunication Management Network, and Telecommunications Information Networking Architecture.

In order to fulfill the requirements of the next generation telecommunications systems, the database architecture must be fault tolerant and support real-time transactions with explicit deadlines. The internals of the RODAIN DBMS are designed to meet the requirements of telecommunications applications including a real-time access to data, fault tolerance, distribution, object orientation, efficiency, flexibility, multiple interfaces, and compatibility with telecommunications practices. The requirements are, to some extent, conflicting. Therefore, the RODAIN database system is based on trade-offs; novel and innovative solutions are used only when tried and tested methods are found to be insufficient.

References

1. R. Abbott and H. Garcia-Molina. Scheduling real-time transactions: A performance evaluation. *ACM Transactions on Database Systems*, 17(3):513-560, September 1992.
2. I. Ahn. Database issues in telecommunications network management. *ACM SIGMOD Record*, 23(2):37-43, 1994.
3. M. Appeldorn, R. Kung and R. Saracco. Tmn + in = tina. *IEEE Communications Magazine*, 31(3):78-85, March 1993
4. R.F.M. Aranha, V. Ganti, S. Narayanam, C.R. Muthukrishnan, S.T.S. Prasad and K. Ramamritham. Implementation of a real-time database system. *Information Systems*, 21(1):55-74, 1996.
5. T.F. Bowen, G. Gopal, G. Herman and W. Mansfield Jr. A scale database architecture for network services. *IEEE Communications Magazine*, 29(1):52-59, January 1991.

6. S. Cha, B. Park, S. Lee, S. Song, J. Park, J. Lee, S. Park, D. Hur and G. Kim. Object-oriented design of main-memory dbms for real-time applications. In *2nd Int. Workshop on Real-Time Computing Systems and Applications*, pages 109-115, Tokyo, Japan, October 1995. IEEE Communication Society.

7. K.P. Eswaran, J.N. Gray, R.A. Lorie and I.L. Traiger. The notions of consistency and predicate locks in a database system. *Communications of the ACM*, 19(11):624-633, November 1976.

8. J.J. Garrahan, P.A. Russo, K. Kitami and R. Kung. Intelligent network overview. *IEEE Communications Magazine*, 31(3):30-36, March 1993.

9. M.H. Graham. Issues in real-time data management. *The Journal of Real-Time Systems*, 4:185-202, 1992.

10. M.H. Graham. How to get serializability for real-time transactions without having to pay for it. In *Real-time System Symposium*, pages 56-65, 1993.

11. J.R. Haritsa, M. Livny and M.J. Carey. Earliest deadline scheduling for real-time database systems. In *Proceedings of the 12th Real-Time Symposium*, pages 232-242, Los Alamitos, Calif., 1991. IEEE, IEEE Computer Society Press.

12. Yennun Huang and Chandra Kintala. Software implemented fault tolerance: Technologies and experience. In *The 23rd International Symposium on Fault-Tolerant Computing*, pages 2-9. IEEE, 1993.

13. Svein-Olaf Hvasshovd, Øystein Torbjørnsen, Svein Erik Bratsberg and Per Holager. The ClustRa telecom database: High availability, high throughput, and real-time reponse. In *Proc. of the 21st VLDB Conference*, pages 469-477, San Mateo, Calif., 1995. Morgan Kaufmann.

14. ITU. *Common Management Information Protocol specification for CCITT applications. Recommendation X.711*. ITU, International Telecommunications Union, Geneva, Switzerland, 1991-2.

15. ITU. *Principles for a Telecommunications Management Network. Recommendation M.3010*. ITU, International Telecommunications Union, Geneva, Switzerland, 1992.

16. ITU. *Distributed Functional Plane for Intelligent Network CS-1. Recommendation Q.1214*. ITU, International Telecommunications Union, Geneva, Switzerland, 1994.

17. ITU. *Global Functional Plane for Intelligent Network CS-1. Recommendation Q.1213*. ITU, International Telecommunications Union, Geneva, Switzerland, 1994.

18. ITU. *Intelligent Network Distributed Functional Plan Architecture. Recommendation Q.1204*. ITU, International Telecommunications Union, Geneva, Switzerland, 1994.

19. ITU. *Draft Q.1224 Recommendation IN CS-2 DFP Architecture*. ITU, International Telecommunications Union, Geneva, Switzerland, 1996.

20. H.V. Jagadish, D. Lieuwen, R. Rastogi, Avi Silberschatz and S. Sudarshan. Dali: A high performance main memory storage manager. In em Proceedings of the 20th VLDB Conference, pages 48-59, 1994.

21. Young-Kuk Kim and Sang H. Son. Developing a real-time database: The StarBase experience. In A. Bestavros, K. Lin and S. Son, editors, *Real-Time Database Systems: Issues and Applications*, pages 305-324, Boston, Mass., 1997. Kluwer.

22. Y. Kiriha. Real-time database experiences in network management application. Tech. Report CS-TR-95-1555, Stanford University, USA, 1995.

23. J. Kiviniemi, T. Niklander, P. Porkka and K. Raatikainen. Transaction processing in the RODAIN real-time database system. In A. Bestavros and V. Fay-Wolfe, editors, *Real-Time Database and Information Systems*, pages 355-375, London 1997. Kluwer Academic Publishers.

24. T. Kuo and A.K. Mok. Application semantics and concurrency control of real-time data-insensive applications. In *Proc. of Real-Time System Symposium*, pages 76-86, 1993.

25. M. Lehr, Y.-K. Kim and S. Son. Managing contention and timing constraints in real-time database system. In *Proceedings of 16th IEEE Real-Time Systems Symposium*, Pisa, Italy, December 1995.

26. K.-J. Lin. Consistency issues in real-time database systems. In *Proc. of the 22nd Hawaii Int. Conf. On System Sciences*, pages 654-661, 1989.

27. C.L. Liu and J.W. Layland. Scheduling algorithms for multiprogramming in a hard real-time environment. *Journal of the ACM*, 20(1):46-61, January 1973.

28. OMG. *CORBA: Common Object Request Broker Architecture and Specification.* Number 91.12.1. Revision 2.0 in OMG Document. John Wiley & Sons, New York, N.Y., 1996.

29. OMG. *The Common Object Request Broker: Architecture and Specification.* Number Revision 2.2 in OMG document. John Wiley & Sons, New York, N.Y., 1998.

30. OMG. *CORBAservices: Common Object Service Specification.* Number December 1998 in OMG Document. John Wiley & Sons, New York. N.Y., 1998.

31. P. Porkka and K. Raatikainen. CORBA access to telecommunications databases. In D. Gaïti, editor, *Intelligent Networks and Intelligence in Networks*, pages 281-300, Paris, France, 1997. Chapman & Hall.

32. Dick Pountain. The Chorus microkernel. *Byte*, pages 131-138, January 1994.

33. K. Raatikainen. Real-time databases in telecommunications. In A. Bestavros, K.-J. Lin and S.H. Son, editors, *Real-Time Database Systems: Issues and Applications*, pages 93-98. Kluwer, 1997.

34. K. Raatikainen, T. Karttunen, O. Martikainen and J. Taina. Evaluation of database architectures for intelligent networks. In *Proc. of the 7th World Telecommunication Forum (Telecom 95), Technology Summit, Volume 2*, pages 549-553, Geneva, Switzerland, September 1995. ITU.

35. K. Raatikainen and J. Taina. Design issues in database systems for telecommunication services. Report C-1995-16, University of Helsinki, Dept. of Computer Science, Helsinki, Finland, September 1995.

36. K. Ramamritham and C. Pu. A formal characterization of epsilon serializability. *IEEE Transactions on Knowledge and Data Engineering*, 7(6), December 1996.

37. L. Sha, J.P. Lehoczky and E.D. Jensen. Modular concurrency control and failure recovery. *IEEE Transactions on Computers*, 37(2):146-159, February 1988.

38. J. Siegel, editor. *CORBA Fundamentals and Programming.* John Wiley & Sons, New York, N.Y., 1996.

39. J.A. Stankovic and S.H. Son. Architecture and object model for distributed object-oriented real-time databases. In *IEEE Symposium on Object-Oriented Real-Time Distributed Computing (ISORC'98)*, pages 414-424, Kyoto, Japan, April 1998.

40. J.A. Stankovic and W. Zhao. On real-time transactions. *ACM SIGMOD Record*, 17(1):4-18, March 1988.

41. J. Taina and K. Raatikainen. Experimental real-time object-oriented database architecture for intelligent networks. *Engineering Intelligent Systems*, 4(3):57-63, September 1996.

42. J. Taina and K. Raatikainen. Database usage and requirements in intelligent networks. In D. Gaïti, editor, *Intelligent Networks and Intelligence in Networks*, pages 261-280, Paris, France, 1997. Chapman & Hall.

43. Øystein Torbjørnsen, Svein-Olaf Hvasshovd and Young-Kuk Kim. Towards real-time performance in a scalable, continuously available telecom DBMS. In *Proc. of the First Int. Workshop on Real-Time Databases*, pages 22-29. Morgan Kaufmann, 1996. http://www.eng.uci.edu/ece/rtdb/rtdb96.html.
44. V. Wolfe, L. DiPippo, J. Prichard, J. Peckham and P. Fortier. The design of real-time extensions to the open object-oriented database system. Technical report TR-94-236, University of Rhode Island, Department of Computer Science and Statistics, February 1994.
45. P.S. Yu, K.-L. Wu, K.-J. Lin and S.H. Son. On real-time databases: Concurrency control and scheduling. *Proceedings of the IEEE*, 82(1):140-157, January 1994.

Overview of Data Management Issues and Experiments in TINA Networks

Yann Lepetit

France Telecom/CNET, 2 Avenue Pierre Marzin, 22300 Lannion, FRANCE
yann.lepetit@cnet.francetelecom.fr

Abstract. This paper presents an overview of data management issues for TINA platforms. TINA networks increase the need to integrate in a more seamless framework data management, distributed computing and management facilities to provide in a dynamic and flexible way new telecom services. To serve this purpose, FT/CNET has conducted, since the beginning of TINA, architectural studies on the data aspects in the TINA framework and has co-operated with both industrial and academic partners to different experiments in this field. The first experiment is a persistence and query service strongly based on OMG specifications but with optimised extensions to match TINA needs The second experiment is a CMIS like TINA management service strongly based on federated DBMS technology. This paper presents a balance sheet of all this approach.

1 Introduction

Since the beginning of eighties some network operators [1] [2] were aware of data management issues in telecom switches. During the nineties, the telecom research community was strongly influenced by the distributed object technology (Open Distributed Processing, Object Management Group), mainly through the TINA [3] project which proposes an architecture for future telecommunication networks. But, if we look at the application field of TINA, we can see that databases are of big importance. Within Intelligent Network, more and more databases are needed to store service information (e.g. subscription information, charging information, re-routing information, statistic information, etc), leading some Service Control Points to be mainly database servers. New Internet Mediation platforms, acting as gateways between pure telecommunication networks and the web/internet network, also require powerful data management facilities. But the data requirements are not conventional in this field : (1) Telecom is not a data intensive field and telecom people do not want to invest too more in database methodology and technology. (2) Telecom databases are usually large with real time constraints, with not too complex schema. (3) Telecom networks have very complex distribution requirements based on different views over data; the telecom service view which is rather simple and the management view (with the Telecom Management Network sense) which is more complex and distributed.

W. Jonker (Ed.): Databases in Telecommunications, LNCS 1819, pp. 16–27, 2000.

The management view is usually a high level view which requires the federation of distributed objects and databases. Strong tendency is also to use specific standards to access distributed data, close to data management but different: CMIS (Common Management Information Service)[12], X500/ladp (directory standards) [14].

As a consequence, data design methodology, data distribution and data interfaces remain to a large extent specifics in the telecom field. To cope with that, we have investigated these issues in a TINA/OMG framework, with industrial and academic partners:

- We have proposed for the TINA project, a data architectural framework [4] based on the ISO ODP standard [5] which is the overall design methodology chosen by TINA.
- We have designed with Ardent/O2 technology, TINA oriented data services [7] for a TINA/OMG platform within the European ACTS ReTINA project, in order to provide a telecom service view on data. We have evaluated these data services with an extended OO7 benchmark.
- We have designed with University of Versailles, a CMIS like service [11] for TINA/OMG platform, in order to provide a management oriented view on data.

This paper presents an overview of these works, discusses some experiment results and concludes.

2   TINA Data Management Framework

The objective is to provide an architectural framework to cope with both management (with the TMN sense) and data issues during the design phase of TINA Networks [4], as these two fields are closely related. This framework offers a methodology to express distributed data requirements at conceptual and logical levels. It is strongly based on the ODP standard [5], which is well accepted by telecom and TINA communities.

2.1   Recall on ODP Framework

The ODP reference model is an ISO standard very useful to cope with distribution provided by OMG products (or RMI, DCOM). It proposes different abstract levels to define and design a distributed system (a telecom network or platform in our case):

- The information viewpoint focuses on the conceptual aspect of a system, without bearing in mind the realisation of this system. The semantic of application is defined by means of information objects.
- The computational viewpoint represents distributed applications as computational objects defined with an Interface Definition Language (IDL) independent of their programming language. A computational object is a distribution unit which is independent of its physical distribution.

– The engineering and technological viewpoints focus on programming language, middleware and product developments. Mapping of objects onto physical sites is decided at this level.

These different levels of abstraction are close to the conventional ones find in the database field: conceptual level, logical level, physical level. The data management framework integrates these database abstraction levels into the ODP framework and copes with telecom data requirements presented in the introduction. The main focus is on information and computational viewpoints of distributed data, engineering and technological viewpoints being investigated by the ReTINA and TINA/CMIS experiments.

2.2   Information Viewpoint

Extending slightly the ODP reference model, we group certain ODP information objects into information bases. This information base concept comes from the telecom TMN field (MIB: Management Information Base). But, we refined it to model both the management view on Managed Objects (with their included data) and the telecom service view on data, as lot of data are shared by these two types of applications. In a conventional database approach, a service information base is a conceptual schema modelling objects which are stored in a database. An information base represents a set of information objects defined by the same designers and naming authority: Objects pertaining to a management information base named with a hierarchic naming tree, or objects pertaining to a service information base named within a flat naming space for example. No restriction is placed on the composition of an information base, and a global management information base can be composed of a federation of local service information bases for example. The semantics of an information base is defined with the OMT/UML graphical language and refined, if needed, with the GDMO specification language (Guideline for Definition of Managed Objects)[6], corresponding to TINA choices for information modelling languages.

In order to illustrate information base, a simplified example is given hereafter. It describes a management information base modelling information used by telecom services to process user calls. In that example, several services are provided to users (phone calls, internet accesses, credit card); specific terminal features and accounts are associated to them.

2.3   Computational Viewpoint

At the computational level, the information bases are mapped into objects bases. An object base groups objects that can be accessed with the same kind of computational interface and located at the engineering level on the same corba server. The various interfaces that can be used are: persistence service interfaces, query service interfaces, management interfaces.

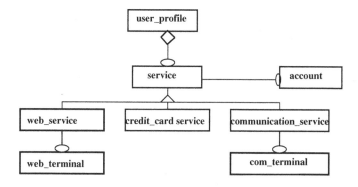

Fig. 1. Information Base represented with UML

An object base can be close to database concept, when its objects are stored in a DBMS. But the concept is enlarged to objects that are not necessarily managed by a conventional DBMS: Managed objects for example that can be stand alone corba objects running on a permanent server. These objects can be retrieved with CMIS filters that are similar to query language predicates.

Object bases are usually mapped onto servers. But a global view over several servers is sometime needed. This is the case for CMIS, where the query scope is a whole platform encompassing several servers (typically one server for a DBMS and several severs for stand alone objects). We have then local object bases which must be federated into a global object base.

Following the conventional database approach, object bases are defined with object base definition language (logical schema definition). In fact, stretching the ODP rules, we may have several languages depending of the engineering tools proposed by relational and object DBMS providers : a pure IDL (OMG Interface Definition Language), an extended IDL (some tools generate SQL/Data Definition Language or ODMG/Object Definition Language from extended IDL definitions), or ODMG/ODL definitions (some tools generate IDL from ODL).

To sum up, the computational level allows to integrate the management and data distribution choices in a way as much as possible independent of the engineering and technological choices.

3   ReTINA Experiment

ReTINA is an European ACTS project (96-99) which aims the design and the implementation of an industrial Distributed Processing Environment based on corba and matching the TINA needs. Main participants are telecom operators, telecom product providers, ORB and DBMS providers.

Within this project FT and Ardent/O2 Technology have co-designed a persistence and a query service [7] implemented by O2 technology on Orbix and Cool

ORBs on top of O2 DBMS. The main objective is to realise data services integrating TINA requirements while keeping compatibility with OMG standards and providing good performances. These services are intended to provide "telecom service view" on data in the TINA architecture, the "management view" being provided by CMIS like services presented in the next section.

3.1   Persistence Service

In TINA service specifications, data are defined as persistent objects like: user profile, service profile, terminal profile, accounting data, etc. The persistence service is mainly used to provide a telecom service view on data. Our starting point to provide this view was the OMG persistence service [8] which is a complex service that proposes to manage compatibility with main database standards by means of several "protocols" (or profiles):

– The Direct Access (DA) profile proposes to define persistent objects with a subset of OMG/IDL (only attributes). This profile is mainly proposed for relational DBMS. A persistent object is identified by a Persistent Identifier object (PID), allowing to mask the DBMS physical identifier. But as main DBMS products provide logical identifiers, this PID is frequently redundant. During its life time, an object can be persistent or transient The persistence is explicitly and dynamically managed with connect/disconnect operations provided by a Persistent Object Manager object. This feature is not really necessary for telecom applications. A Persistent Data Service object models, in the Corba world, the pure DBMS programming interfaces (open and close database, open and close session, lookup data by oid, etc).
– The Dynamic Data Object profile is a very basic profile where each data must be explicitly encoded (name, value, type, size). It is not really user friendly, its only advantage is to be neutral.
– The ODMG profile is proposed for ODMG conformant DBMS , but there is very few and imprecise OMG recommendations for this profile.

Within TINA, telecom services see their data as persistent objects and expect to manipulate them with as much as possible "persistence transparency". As a consequence, ReTINA try to simplify and to clarify the persistence service implementation. To stay conformant to the OMG standard, ReTINA has implemented a pure OMG/DA profile with some optimisation (storage of the PID in the persistent object in order to save disk IO for example). But, as this DA profile is rather complex for TINA needs and not really seamless, we added an ODMG profile.
This ODMG profile provides a more transparent Corba object persistence. Persistent objects are defined with all the capabilities of OMG/IDL (attributes and operations). They are statically persistent. They are looked up with their logical ODMG name and their state are transparently saved at the end of transactions. Moreover, we added very useful features which were not in the OMG standard:

- A simplified persistence service binding: a unique "init" operation is provided to mask several ODMG initialisation operations: open session, open database, begin a transaction. A one to one mapping is chosen for the persistence corba server and database (same Corba server name and database name).
- Three transactional modes, chosen at binding time: A user driven mode, where a Corba client controls via IDL interfaces, the transaction operations provided by the DBMS (begin, abort, commit,.). A per operation mode, where each operation made by a client is transparently transactional. A monitored mode, handled by a transactional monitor (OTS), this mode allows to perform distributed transactions, but is time consuming.

3.2 Query Service

Within TINA, the main purpose of the query service is to extract state information from connection graphs (SDH, ATM, routing graph, numbering graph) which are intensively used in the communication service part of the TINA architecture.

Our starting point was the OMG Query Service (OQS) [9] which provides predicate-based operations on collections of objects. Queries are expressed with a DBMS query language which can be either OQL from ODMG or SQL from ISO. In a pure OMG approach these query languages should act directly on transient or persistent Corba objects which may not always rely on DBMS to interpret the queries ("vanilla OMG objects with no extra capability"). A Query Evaluator provides a generic evaluate operation which interprets the query. Collections provide Corba interfaces to manipulate set, list or bag of elements. The OMG collections only manipulate non-typed elements (any). The result of a query can be an non-typed element (any) or a Corba object reference. Numerous interactions are usually required to get the final result: query returning a collection reference, collection iterator creation, iteration, access to a result object.

Such a query service acting on transient objects is too complex to implement and can lead to very poor performances. Then, we have designed an OQL query service acting only on persistent objects stored in an object DBMS and relying strongly on the DBMS evaluator. A lightweight profile is added to better match the "query transparency" and performances waited by TINA people. We mainly add the following features:

- Manipulation of ODMG collections via IDL interfaces allowing to manipulate typed elements.
- Use of IDL sequences to transmit back the result of queries.
- Optimised update operations compatible with OQL (insert_element, remove_element, remove_selected_element). These operations allow to directly update a collection without using a collection factory.
- Explicit lock modes on objects in order to decrease the concurrency conflicts. These lock modes are managed by the underlying DBMS. The reuse of the OMG concurrency service for that purpose was not possible due to the lack of links between the OMG locking and DBMS locking mechanisms.

3.3 Persistence Service and Query Service Evaluations

The performances of the persistence and query services have been evaluated with an extended OO7 benchmark [10]. The performances of ReTINA three tier architecture was also compared to pure O2 DBMS client server architecture, in order to evaluate the overhead brings by the Corba distribution.

The persistence service was evaluated with various types of navigation between objects, these navigations being implemented as methods of the objects stored in the database. For a read navigation the ReTINA overhead is negligible (1.5 per cent). For an update navigation, the difference is more significant, depending of the number of data being updated during the navigation (3 per cent). For the scan of big strings, the ReTINA overhead is important (ranging from twice to ten times) due to the numerous Corba interactions needed for that operation.

The query services was evaluated with graph queries of different selectivities: exact match, 1%, 10%, all. The results show that the ReTINA overhead increases quickly with the decrease of the selectivity (ranging from 75 percent to 600 percent). One possible explanation is that Corba processing is linear with the number of results, where DBMS processing is optimised for large amount of data (caches, indexes, etc). This effect is greatly attenuated by the lightweight profile (response time divided by 3 in the most favourable case) which requires less corba interactions thanks to sequence transmission.

A multiple Corba clients trial shown the interest of three tiers architecture which allows to parallelise the application processing and the DBMS upper layer processing. This is particularly true for the query service, where the path traversal is performed by the DBMS. But this may lead to a Corba server/DBMS client bottleneck with the persistence service, in case of thin corba client implementation.

4 TINA CMIS Experiment

The main objective of this experiment is to provide " TMN agent like " facilities for the management of TINA platforms based on CORBA [11]. To deal with these issues, we have designed a TINA/CMIS Query Service. It provides a real query language, called CMIS-L, compatible with CMIS/CMIP [12]; and dynamic schema modification possibilities. This service provides a high level management view on heterogeneous objects like stand alone Corba objects, persistent objects managed by a persistence service or a query service.

The originality of the approach is that it particularly capitalises on database query and database federation technologies, but it applies them outside the database context, in the corba context.

4.1 Managed Object Model and Naming

To provide interoperability between telecom service platforms and management platforms, a common management model is needed. For this purpose, at the beginning of the nineties ISO/TMN has defined it's own object model GDMO to model Managed Objects. At this time, object languages and distributed object technologies were not used in the telecom field and a specific object model was needed. This is no longer true, particularly in TINA/Corba architecture. Then, our Managed Objects are defined with the IDL/OMG interfaces. These interfaces offer transparency with engineering representation (relational data, object data, files, programming languages) and physical localisation of Managed Objects.

Within TMN, Managed Objects are organised into a specific Naming Tree (NT), which models a containment tree. As the concept of naming tree doesn't exist in Corba, we propose to map it with an IDL-like syntax as described hereafter.

Naming_tree User_Information_nt
root User_Information_root {
    Interface User_profile   **Interface** Terminal
    named_by User_Information_root   **named_by** Service
    with user_rdn ;   **with** terminal_rdn ;
    Interface Service   **Interface** Account
    named_by User_profile   **named_by** Service
    with service_rdn ;   **with** account_rdn ;
                                     } ;

4.2   Object Base Federation

On a telecom platform, services are usually designed and developed in an autonomous way by different development teams. New services, co-existing with previous services, are deployed during the life time of the platform. A global management view must be permanently provided on these heterogeneous services. We are then typically faced to a federation issue, but related to managed objects and not only databases. This federated management view is provided by a global Object Base (OB) which is a " virtual base " composed of a federation of local Object Bases. A local OB is a group of objects located in a Corba server. These objects are supported by heterogeneous servers : (1) pure Corba server providing no query and DBMS facilities, (2) Corba server with a persistence/query service relying on a relational DBMS, (3) Corba server with a persistence/query service relying on an object DBMS.

As objects bases are designed in an autonomous way, each Corba server exports some local schema information to the CMIS Query Service in order to build a global management view of these heterogeneous object bases: the name of the server (e.g. object base), its type related to its query capabilities, the part of the global naming tree it implements, and the part of the global object base it implements (its implemented interfaces).

4.3   Querying Facilities

In current telecom networks, objects are mainly "queried" and created by means of protocols like CMIP/CMIS (in a way which can be compared, in the database field, to a direct use of RDA [13] instead of SQL). This approach, too closely linked to a specific " protocol view " (ASN1 encoding), presents a lot of drawbacks: non user friendly interfaces, impedance mismatch with corba IDL, etc. Moreover, SQL or OQL are often perceived as too complex by the telecom community and they can not be used directly for management purposes: SQL don't manipulate real objects and OQL don't allow to create or delete real objects. As a consequence, we propose a real query language called CMIS-L, which keeps the semantic of CMIS, but can be transmitted easily by corba (or http).

If CMIS-L has some syntax similarities with database query languages, its semantic is different and it allows to remotely create and delete real objects. CMIS-L provides to users a high level query language to query, with filters, objects named by a naming tree. Filters are very similar to assertionnal predicates of database query languages. CMIS-L offers many advantages : (1) Simplicity of querying, a CMIS-L query is a string expression concise and easy to express. (2) Transparency of distribution, CMIS-L queries are evaluated upon the global OB (federation aspects are assumed by the TINA Query Service). Finally (3) efficiency, CMIS-L query processing is based on a cost model (like query optimiser of DBMS's, but implementing a new algorithm, as the CMIS query model is different from the SQL or OQL query model), this cost model is used to find the best execution plan to retrieve both data and stand alone objects.

CMIS-L offers operations to create, delete, query and update one or more objects, as well as to initiate actions on them. The operations are performed on a hierarchy of objects defined by the containment tree.

An example of a CMIS-L query is given hereafter.

GET u.name, u.user_id, s.service_id
SCOPE DN = {} **LEVEL** = 1
FROM User_Profiles(u), Services(s)
FILTER u.name like 'A*'

This query searches services of a subset of user Managed Objects. The subset pertains to a sub-tree identified with the "scope" and "level" clauses. Among these objects we kept those pertaining to the User_Profiles and Services classes and selected by the filter clause.

Some CMIS extensions are provided in order to avoid ambiguities sometimes present in pure CMIS, but an upward compatibility with pure CMIS is kept. One extension, for example, allows to retrieve path expressions. This permits to keep the containment links in the query result. This was not possible in pure CMIS in which only set of objects can be retrieved, with loss of the containment semantic. As, the result of a query can be complex, some browsing facilities are also provided to explore the tree like structure of the result. A more complete presentation of CMIS-L and of its semantic can be find in [15].

4.4   CMIS Query Service

CMIS -L is provided by a CMIS Query Service component located on a telecom services platform : an ORB supporting different telecom service objects and their associated data. It provides a common management view on federated corba servers. The technological heterogeneity and distribution of object managers (pure CORBA server, relational DBMS server, object DBMS server) is masked.

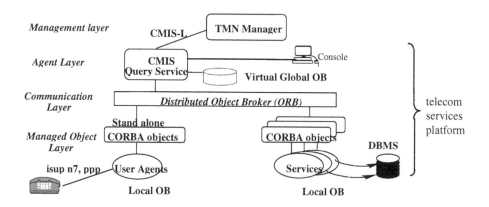

Fig. 2. The TINA/CMIS query service architecture.

The CMIS Query Service is fully integrated in the OMG philosophy. It can co-operate with OMG data services : persistence service, query service, transaction service. Its implementation relies as much as possible on services provided by CORBA: naming service, interface repository and dii. However, as it is not possible to store all the needed information in the naming service and the interface repository, a CMIS repository has been developed for object base schema. Queries are addressed to the CMIS Query Service either through an interactive interface, either through a programming interface such as the Evaluate (in string query_CMIS_L, out result) as proposed for OQS. As these queries are expressed upon the global OB, they are decomposed by the CMIS Query Service accordingly to the different local object base and Corba server capabilities. For a corba server built upon a RDBMS or OODBMS, sub-queries are delegated to the server and sent to the DBMS which evaluates efficiently complex queries on a large amount of data. For a pure corba server in charge of stand alone objects, retrieve is done one object at a time. Of course, the performances are completely different for these different solutions.
This CMIS query service is under implementation by University of Versailles on an OrbixWeb platform.

5   Conclusions

The analyses performed for the TINA Data Management Framework were very useful to understand data distribution issues in telecom networks. Moreover, this framework is enough abstract to be used for new technology like Enterprise Java Beans (EJB). It allows to distinguish entity EJBs from session EJBs during the design phase of an application for example. But there is also some disappointments, as we thought within TINA, that convergence in object technologies will lead to standard OMG service implementations with well defined interfaces, and that convergence in database technology will provide a common query language merging OQL and SQL3. In that case, the computational viewpoint would have been very independent from the engineering viewpoint. We have in fact greatly underestimated the competition between software product providers (and the economical stacks), leading them to differentiate their offer and to try to impose their own standard. We have still two families of DBMS products: relational and object, and if we look at Corba persistence services, we have almost as many different interfaces as DBMS and ORB providers.

ReTINA experiment provides an industrial implementation of data management services keeping as much as possible conformance with OMG standard, with enhancements for TINA needs. It demonstrates the feasibility of OMG conformant persistence and query services. The query service seems to be one of the few implementation to exist and its lightweight profile provides good performances on rather big databases. Main optimisations are implemented at the Corba level with no impact on the DBMS product. This is important if we want to benefit of the evolutions of DBMS standards and products. The three tiers architecture seems to be very promising to solve data performance issues in telecommunication, specially if it is coupled with an efficient cache mechanism (like the one provided by O2 DBMS). Main results of this experiment are now available as a commercial product and can be used by telecom product providers and telecom operators.

Regarding the management of telecom platforms, telecom and OMG people have spent a lot of time on the GDMO IDL translation problems. But is it the main problem if we use distributed object technology everywhere? Due to telecom deregulation, new telecom services must be designed and deployed very quickly, once deployed they must be modified on line to better match the customer needs. As a consequence, telecom management must become more dynamic: objects must be dynamically modified, new objects must be introduced with little impacts on the management facilities. In this context, the data management contribution has been very little studied. The on-going implementation of our TINA/CMIS Query Service demonstrates that the use of database concepts and techniques to implement management facilities close to telecom management standard is very profitable. It contributes to provide important features like query optimisation, dynamic schema evolution, dynamic federation which are of main importance for telecom operators.

Acknowledgements. The author would like to thank T. Delot, B. Finance, G. Gourmelen, S. Habert for their participations in the works presented in this paper.

References

1. A.V. Aho and als "*Database systems*" The Bell System Technical Journal volume 61 November 1982
2. B. Kerherv, Y. Gicquel, G. Legac, Y. Lepetit, G. Nicaud, "*Sabina-rt : a distributed DBMS for telecommunication*" First Extending Data Base Technology conference Venice March 1988
3. F. Dupuy, G. Nilsson, Y. Inoue, "*The TINA Consortium : Towards Networking Telecommunications Information Services*" International Switching Symposium, Berlin April 95
4. Y. Lepetit, "*Overview of a Data Management Framework for TINA*" , TINA'96 Conference, Heidelberg, September 1996
5. ISO/IEC JTC1/SC 21 n 7524, "*Reference model of ODP Part 2 : Descriptive Model*", February 1994
6. ISO/IEC 10165-4 "*Guideline for the definition of the Managed Objects : GDMO*", 1992
7. S. Habert, Y. Lepetit ReTINA ACTS project deliverable RT/TR-97-12.0.2-*WP2* "*Specifications of persistence and query services V2*" June 1997
8. Object Management Group, "*Persistence Service Specifications*" OMG doc 95-3-31 March 1995
9. Object Management Group, "*Object Query Service specifications*", Itasca, Objectivity, Ontos, O2, Servio, Sunsoft, Sybase, Taligent March 1996
10. G. Gourmelen, Y. Lepetit ReTINA ACTS technical report TR-98-07 "*Benchmark of persistence and query services*" August 1998
11. J. Fessy, B. Finance, Y. Lepetit, P. Pucheral, "*Data Management Framework & Telecom Query Service for TINA*", fifth International Conference on Telecommunication Systems Modelling and analysis, Nashville, March 1997
12. ISO/IEC 9595 and CCITT/X710, "*Common Management Information Service : CMIS*", 1992
13. ISO/IEC/SC21/7689 RDA : "*Remote Database Access*", 1993
14. ISO/IEC/9594 and CCITT/X501 : "*The Directory model*", 1990
15. T. Delot, B. Finance, Y. Lepetit, A. Ridaoui "*TINA Service Platform Management Facilities*" Networks and Services Management conference GRES99, Montreal, June 1999

Database Architecture for Location and Trajectory Management in Telecommunications

Sang K. Cha, Kihong Kim, and Juchang Lee

KINS (Knowledge and Information Net for Sharing) Laboratory
School of Electrical Engineering
Seoul National University
{chask, next, juch }@kdb.snu.ac.kr

Abstract. The need for managing the location and the trajectory of mobile terminals efficiently rises rapidly as all the wireless carriers in the United States implement the location measurement function for the E-911 service. On the other hand, the main-memory DBMS is being taken as an economically viable alternative to the disk-based DBMS as tens or hundreds of gigabytes of memory becomes no longer a luxury in the computer systems of near future.

In this paper, we bring our past experience of developing a main-memory storage system called XMAS and its spatial extension XMAS-SX in the context of the location and trajectory management in telecommunications. Our experiment tells that the main-memory database is superior to the disk-based OODBMS, especially when the workload is highly skewed toward random object updates. After the presentation of the current implementations of XMAS and XMAS-SX, we present desirable five layers of the embedded location and trajectory management database system, and discuss the efficient indexing of the trajectory information for query processing.

1 Introduction

With tens or hundreds of gigabytes of memory becoming no longer luxurious in the computer systems of near future, the main-memory DBMS is considered as an economically viable alternative to the disk-based DBMS, especially, in the embedded system applications demanding high performance. Telecommunication, including the Internet applications in the broad sense, is representative of such applications that can take advantage of the main-memory database for its fast and relatively even response time and high throughput rate. The recent products in the main-memory DBMS market aim at various telecommunication applications for this reason [1].

On the other hand, in telecommunications, the wireless mobile communication has become popular. This leads to the increasing demand for managing the location and trajectory information of mobile terminals. In many of the current mobile communication systems, the HLR/VLR database records the IDs of cells or location areas where the mobile terminal belongs. However, the technologies

W. Jonker (Ed.): Databases in Telecommunications, LNCS 1819, pp. 174–190, 2000.

measuring a more precise location based on the cellular network structure or the GPS (Global Positioning System) are already available [2, 3]. Especially, the precision and the economic viability of the network-based radiolocation are expected to improve further because the United States FCC requires all the wireless carriers to be capable of reporting mobile terminal locations with an accuracy of 125 m by October 1, 2002.

Once the radiolocation becomes part of the wireless network, many types of value-added services are conceivable, such as location-sensitive billing and direction-sensitive message broadcasting. Most of these services involve spatial or spatio-temporal queries to the dynamic embedded location and trajectory management database. This type of database has been already demanded by some telecommunication applications, such as the AVL (automatic vehicle location) domain of ITS (intelligent transportation system) [4]. For example, the vehicles in the delivery and pick-up service report their location updates to the management center through the wireless network, and the center makes decision on which vehicle to dispatch upon the customer request. With the location management part of the network, the implementation of ITS can be made simple.

This paper is concerned with the implementation of the location and trajectory management database based on the main-memory architecture. We propose to employ this main-memory architecture based on our prior experience of developing a general-purpose main-memory storage system XMAS [5, 6] and its extension called XMAS-SX for spatial data handling [7]. This choice of the main-memory database is first justified based on the fact that telecommunication systems are in general soft real-time systems that require fast and even response time. In addition, the location management involves massive amount of location updates and requires a high rate of update transaction processing. In general, the main-memory database shows a higher throughput of updates than the disk-based one because the updates to the main-memory database lead to sequential disk writes for recording logs while the updates to the disk-based database lead to random disk writes. In this paper, we present an architecture that increases the log processing rate of the main-memory database further by incorporating multiple log processing agents.

While we are exploring the use of the main-memory database technology for the location and trajectory management, we do not exclude the disk-based database from the global system architecture. If applications require the whole trajectory to be kept for each mobile object, it is not possible to store all of the trajectory information in main memory because the trajectory database is essentially a historical database that never deletes its entries. In this case, we assume the division of roles between the main-memory and disk-based databases such that the main-memory database serves as the front-end for keeping the recent part of the trajectory and the disk-based database as the back end for keeping the rest of trajectory information.

The contribution of this paper is to bring our past experience of developing the main-memory database to the management of the location and the trajectory of mobile objects as well as relatively stationary geographic informa-

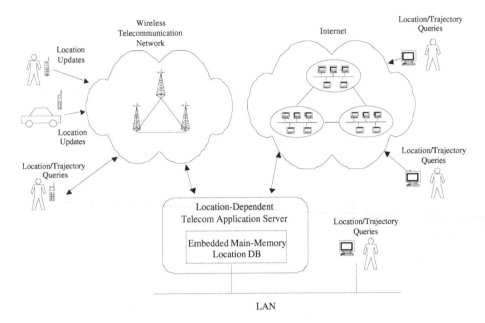

Fig. 1. Location-Dependent Telecom Service Architecture

tion such roads and landmarks. After the discussion of the location-dependent telecommunication applications and the current state of the continuously evolving XMAS implementation, the paper presents an overall architecture of the embedded main-memory DBMS for location management. Finally, we discuss the trajectory indexing, which we take as the key implementation issue in this problem domain.

2   Location and Trajectory in Telecom Applications

Figure 1 shows the problem context that we are interested in. Location updates coming from the wireless network is reflected to the embedded main-memory location database in the application server. The content of this database can be accessed through both the wireless network and the Internet.

Below, we describe some representative location-dependent applications that may be serviced by the above architecture and characterize them in terms of what types of information is needed and whether there is any real-time requirement to keep them in the main memory. For the first, we define the location and the trajectory information below.

2.1   Location and Trajectory Information

– *Timestamped location snapshot* of an object is a triple $<o, l, t>$ representing the most recent location l of an object with its OID o at the reported time

t. The location l is a two-dimensional value representing a coordinate (l_x, l_y). Queries may be posed for all four dimensions. For instance, given an OID, find the most recent reporting time and location of the object, or given a specific region and a certain time interval, find all objects in the region during the time interval. For the efficient processing of the latter type of queries, a three-dimensional index with the location and the time as the key is needed, and for the first type, an additional one-dimensional index with the OID as its key is needed.

– *Location snapshot* of an object is a pair $<o, l>$ representing the most recent location of an object. This is a special case of the timestamped location snapshot with the temporal information dropped off. A two-dimensional spatial index is sufficient instead of the three-dimensional index.

– *Sampled trajectory* of an object is a pair $<o, \{ l_i, t_i \}>$ (where $i = 0, 1, ..., n$) representing the OID o and a timed sequence of locations $\{l_i, t_i\}$. Section 4 further discusses on indexing this type of information for the efficient query processing.

2.2   Characterization of Location-Dependent Telecom Applications

The following list of location-dependent telecommunication applications has been made based on the articles [3] and [8]. While the first focuses on radio-location technologies, it briefly explains many of the applications listed below. The second, although it assumes the GPS technology for obtaining the location information, it presents an interesting picture of the future Internet with location-based addressing, routing, and resource discovery.

1. *Emergency 911 service*: The mobile terminal calls the 911 number, and the 911 service center dispatches one of the standby emergency vehicles near the reported location, and directs the vehicle to one of nearest hospitals. The 911 database records the location of the reported emergency and the trajectory of the dispatched emergency vehicle until its mission is completed. For the real-time k-nearest search ([9, 10]) of emergency vehicles and hospitals, it is desirable to keep the location information of vehicles and hospitals in memory with appropriate spatio-temporal and spatial indexes. Once the mission is completed, the location and trajectory information may be moved from the main-memory database to the disk-based database for archive, except the current location of the emergency vehicle.

2. *AVL and fleet management in ITS*: An individual vehicle continuously updates its location, and the system maintains the trajectory of vehicles. Some applications require the current snapshot of vehicle locations to answer the nearest neighbor query, while others require the trajectory of vehicles. This application is a general case of the E-911 service.

3. *Geographic service querying and advertising*: The current location or the direction of movement of the mobile terminal determines the information service that it receives from the carrier. For example, if the traffic in a certain geographic region is under severe congestion, those approaching the region

receive warning messages. It is necessary to maintain the current location and the planned trajectory of mobile terminals in memory as well as the congested region for the real-time nature of the application.

4. *Location-sensitive billing*: The wireless carrier offers different rates depending on whether the mobile terminal is used at home, in the office, or on the road. The system needs to maintain the user's home and office location in the database to decide the billing rate. If the service includes real-time notification to the user of the billing rate, the memory residence of the home and office location information is desirable.

5. *Fraud detection*: The wireless carrier keeps track of the trajectory of the caller. If a mobile terminal makes two temporally adjacent calls from the geographically distant locations that are impossible to reach during the time interval between two calls, one of them is suspected as fraud. For the on-line fraud detection, it is necessary to keep the timestamped location of the most recent call from the terminal.

6. *Cellular system design and resource management*: By keeping track of the location of mobile terminals, the wireless carrier can utilize its resources better and provide higher quality service. For example, it is possible to configure cells dynamically by keeping track of the number of active mobile terminals in a certain geographic region.

7. *Traffic control system in ITS*: In this ITS domain, it is necessary to maintain the traffic volume information of specific locations such as intersections and road segments to control traffic signs. In this case, the update occurs not on the location but on the value of the objects (intersections or road segments) on the specific location.

The difficulty arises in many of the above applications because each location or trajectory update leads to the corresponding reorganization of the spatial or spatio-temporal indexes. This makes the big difference in the requirement for the location and trajectory management database from the typical geographic information system (GIS) applications such as the cadastral information system where the spatial index update is not frequent.

2.3   OpenGIS Geometry Model

All of the above-mentioned applications require the underlying DBMS to support spatial data types, relationships, and operators. Specifically, we need geometry data types for representing the location and trajectory of mobile objects and the reference geographic information such as the road segments and land marks (lakes, rivers, etc.).

Much work has been done on the representation of spatial data and efficient indexing schemes. In this work, instead of inventing yet another model, we adopt the OpenGIS standard geometry model for the representation of spatial objects [11, 12]. Figure 2 shows the inheritance hierarchy of built-in spatial data types following the core subset of the OpenGIS geometry model. Those surrounded by a bold box such as *LineString* and *Polygon* represent concrete types, which

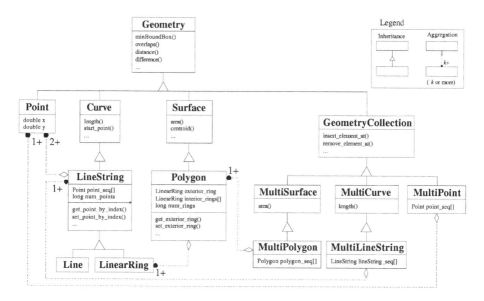

Fig. 2. OpenGIS Geometry Model

can be instantiated. Abstract types such as *Geometry* and *GeometryCollection* only capture the commonality among the subtypes. Additional domain-specific geometry types can be derived from these core types. For instance, the B-spline surface type, which is common in CAD/CAM, can be defined as a subtype of Surface. Derived from the *Geometry* type, all the types in the figure are classified into primitive geometry and geometry collection.

Point, *Curve* and *Surface* are primitive geometry types. Derived from *Curve* is *LineString*, which models a piecewise linear abstraction of curve with a sequence of *Point* objects. *LineString* has two subtypes, *Line* and *LinearRing*. *Line* is a straight segment represented by two end points and *LinearRing* is a piecewise linear abstraction of closed curve. *Polygon* represents two-dimensional objects with area and has *LinearRing* objects to represent its outer and inner boundaries.

The *GeometryCollection* type represents a homogeneous set of geometry objects. It defines methods for insertion, removal, and search of component objects. Three types are derived from this type. *Multipoint* is a concrete type, although *MultiSurface* and *MultiCurve* are abstract types specialized to *MultiLineString* and *MultiPolygon*, respectively.

Table 1 describes the spatial operators defined in [12]. Constructive and metric operators are unary operators except distance. The rest of operators are binary, applicable to any two *Geometry* objects. Among unary operators, the length operator is applicable to the *Curve* and *MultiCurve* types, and the area operator is applicable to the *Surface* and *MultiSurface* types. The relates operator tests the spatial relation between two geometry objects based on the 9-intersection matrix [13].

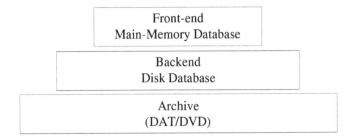

Fig. 3. Multilevel Storage Architecture for Location and Trajectory Management

3   Location and Trajectory Database Architecture

This section presents the multilevel storage architecture for the location and trajectory management, the current state of the XMAS implementation, and an overall architecture of embedded main-memory location and trajectory database that has evolved from it.

Table 1. Spatial Operators

Operator Type	Operator Name
Constructive	copy, boundary, buffer, convex_hull
Relational	equals, disjoint, touches, crosses, within, overlaps, contains, intersects, relates
Metric	distance, length, area Set intersection, union_op, difference, symmetric_difference

3.1   Multilevel Storage Architecture

Figure 3 shows the multilevel storage architecture for the location and trajectory management, where the main-memory database is used for the high-performance access to data. The main-memory database is backed up by the backend disk database because it is not possible to store all of the trajectory information in main memory if applications require the whole trajectory to be kept for each mobile object. The disk database can be again backed up by the archive such as DAT (Digital Audio Tape) and DVD (Digital Versatile Disk).

In this architecture, updates are first reflected in the main-memory database while the reflection of these updates in the disk database is postponed to be processed as a group update. Under this update scheme, a search query is processed in the main-memory database first. If the main-memory database does not have sufficient data for the query, the result is merged with the query result over the disk database.

This multilevel storage architecture gives the following challenges for the efficient cooperation between the main-memory and disk databases.

- *Logging*: Since updates are performed on both main-memory and disk databases, a naive logging scheme will generate two log records for each update redundantly. We expect that the logging scheme of the disk database can be much simplified.
- *Checkpointing*: Checkpointing in main-memory databases is to copy the physical image of memory-resident database to disk for fast recovery from crash. In the proposed architecture, updates in the main-memory database are reflected in the disk database eventually. Thus, if it is possible to reconstruct the main-memory database quickly from the disk database, the checkpointing cost in main-memory database can be saved.
- *Index Recovery*: Updates to indexes don't have to be logged in main-memory databases because indexes can be rebuilt quickly during recovery using bulk-loading algorithms. Bulk-loading in main-memory databases is much faster than that in disk databases because all the records are already in memory. In the proposed architecture, however, since past snapshots of main-memory indexes are kept in the disk database, reconstructing main-memory indexes from the snapshots can be faster than bulk-loading from scratch.
- *Batch Deletion*: It is needed to remove past, inactive records periodically from the main-memory database because the size of memory is limited. Removing records themselves is relatively cheap but the corresponding management of indexes is very expensive. To replace the naive record-by-record removal from indexes, it is desirable to develop a batch-deletion algorithm and an index that facilitates the batch deletion.

3.2   XMAS Implementation

XMAS (eXtensible MAin-memory Storage system) is a main-memory storage system that has been implemented and been evolving at Seoul National University, Korea as the research vehicle for high-performance database applications. Implemented in about 50K lines of C++ code based on the object-oriented software design for the ease of extensibility, it supports the core DBMS functionality such as data persistence, concurrency control, transaction scheduling, and recovery management. XMAS has the following architectural characteristics:

- *Action processing architecture*: The application transaction is defined as a sequence of actions from transaction begin to transaction commit (or abort), and the server takes this action as the unit of processing. A set of built-in actions is provided, and an application may compose its own set of complex actions based on these built-in actions.
- *Multithreaded server architecture*: For the scalable performance over the multiprocessor platforms, the server preallocates a pool of threads called APT (action processing threads) for running the client-requested actions as well as other dedicated system threads for running some asynchronous system functions such as checkpointing and log flushing.

- *Virtual memory database architecture*: The primary database resides in the virtual memory space with a backup copy on the disk. The system provides a set of trusted primitive operations (called actions) for the applications to access the database.

Figure 4 shows the architecture of XMAS with a number of ovals, each of which represents a functional module called manager:

- *Database Manager*: Maintains the memory-resident primary database and thus implements the logical and physical structures of the database. Logically, a database is a collection of containers, each of which is again a collection of fixed or variable-length records. Also implements a library of index classes for the hash table, T-tree [14], R-tree [15], and R*-tree [16].
- *Transaction Manager*: Keeps track of transactions and coordinates other managers to support transaction begin, commit, and abort.
- *Lock Manager*: Implements the container-level two-phase locking for concurrency control [17].
- *Transaction Scheduler*: Supports priority-based transaction scheduling.

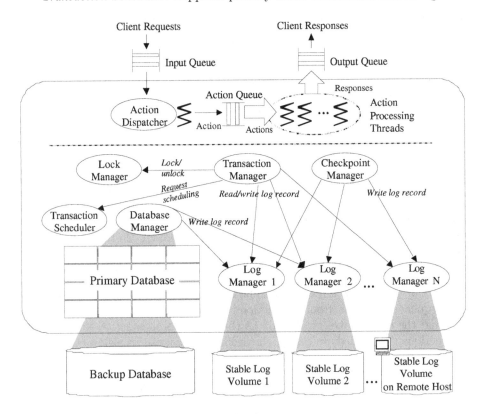

Fig. 4. XMAS Architecture

- *Log Manager*: Manages the in-memory log buffer and flushes it to the stable log devices.
- *Checkpoint Manager*: Implements the fuzzy checkpointing to shorten the restart time [18].
- *Recovery Manager*: Implements the ARIES-based recovery algorithm [?] and manages the restart process after the system crash.

Each of the above managers is implemented as a C++ class encapsulating the core data structures of the XMAS storage system. These classes interact with each other by calling the public interface parts.

One notable change in the current XMAS architecture compared with its previous version ([5–7]) is that it incorporates multiple manager instantiations for certain types of managers. Figure 4 shows the log managers as a representative example. Each log manager has its own logging device such as a disk or a disk combined with a nonvolatile RAM. By instantiating a multiple number of log managers, XMAS increases the processing rate of update transactions in the main-memory database system, where logging is the dominant bottleneck in the system performance. Another new feature in logging is the so-called remote log manager that writes log records on the remote host. This remote log manager is useful for implementing a hot standby main-memory database system for the high availability.

3.3   XMAS Spatial Extension

To meet the high performance requirement in some geographic database applications such as ITS, we have incorporated a spatial data management layer within the XMAS. Named XMAS-SX, it implements a core subset of the OpenGIS geometry model, spatial indexes, and a few spatial query algorithms such as the k-nearest search and the overlapping range query.

Figure 5 shows the comparative performance of XMAS-SX with a comparable spatial database extension module implemented on top of a commercial disk-based OODBMS [20]. Two types of transactions are considered for the synthetic database of 300,000 objects: the range query and the attribute update of 10 randomly selected objects. For the fair comparison of the range query execution time, we ran 1,000 random query executions for the disk-based OODBMS before the measurement to make the database fully cached in the buffer of the OODBMS. Since this disk-based OODBMS caches the data in the application process based on the file server architecture, the tested process architecture is almost identical to that of XMAS-SX. Thus it is not surprising that XMAS-SX is marginally faster than the fully cached disk-based OODBMS in the range query execution time. It should be noted from Figure 5 (c) that XMAS-SX is significantly faster than the disk-based OODBMS by orders of magnitude when the data is not cached.

Figure 5 (b) shows that in the attribute update of 10 random spatial objects, XMAS is faster than the disk-based OODBMS by more than an order of magnitude. This proves that the main-memory database is superior to the disk-based

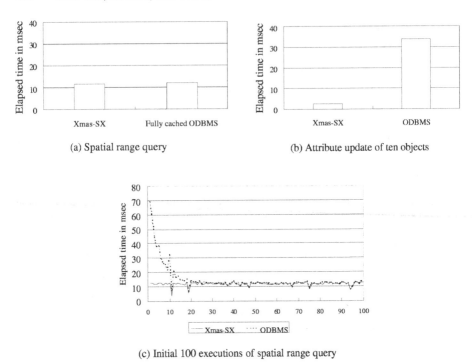

(a) Spatial range query   (b) Attribute update of ten objects

(c) Initial 100 executions of spatial range query

Fig. 5. Comparative Performance of XMAS-SX

OODBMS for the update-intensive load, contrary to the common expectation on the main-memory database needs. The large cache in the disk-based OODBMS does not buy the gain in the update performance.

3.4 Proposed Architecture

Figure 6 shows the five layers of the location-dependent application server architecture.

1. Main-memory storage system
2. ODMG-compliant object model implementation
3. OpenGIS geometry model implementation
4. Location and Trajectory model implementation
5. Application-specific object class implementation

The need for the implementation of the embedded ODMG-compliant object data model has been recognized during the process of developing XMAS-SX. We learned that the development of any object-oriented domain-specific data model like the OpenGIS geometry model eventually needs a general object-oriented model implementation. On top of this general embedded object-oriented data model, the OpenGIS geometry model can be implemented as a library of object

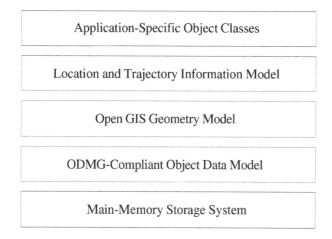

Fig. 6. Location-Dependent Application Server Architecture

classes. Other object-oriented domains such as the GDMO model of TMN [21] can take advantage of this general embedded object-oriented data model. The location and trajectory information model is then implemented on top of the previous three layers. The next section discusses the implementation of this model further focusing on the trajectory information.

At the top level, the application-specific objects are defined. These application objects may be accessed through the object request broker of the CORBA ([22]) from the clients. The CORBA is favored in the telecommunication domain where the heterogeneity resolution is a serious problem. The CORBA provides the transparency at the platform and network levels.

4   Trajectory Indexing

The trajectory of a mobile object is a timed sequence of locations. In terms of the OpenGIS geometry model, it is a 3D *LineString* instance. In this section, we discuss techniques for indexing this trajectory information. As we discussed in section 2, the indexing for the location snapshot and the timestamped location snapshot is relatively straightforward because they essentially represent points in 2D or 3D space. Typical 2D/3D index structures such as R-tree can be used for the efficient processing of range queries and k-nearest queries.

However, treating the trajectory information simply as a 3D object in indexing does not work because the time dimension of the trajectory may extend infinitely and the MBR (minimum bounding rectangle) of the trajectory monotonically grows with time. This makes the index structures useless because of the high ratio of false hits. Another requirement in the context of the main-memory database is that it is desirable to move the old (or inactive) part of trajectories to the disk before the main memory becomes full because it is likely that the recent trajectory information is more frequently accessed than the older one. It is desired for the trajectory index structure to facilitate the movement of its old part without incurring much cost. In the later part of this section, we propose such a scheme based on time-sliced trajectory decomposition.

4.1   Naive Method

One naive way to handle the monotonically growing MBR problem in trajectory indexing is to treat each timestamped location of a trajectory as a separate point entry in the 3D index. Thus for each trajectory, multiple index entries are created one for each location update report. Searching trajectories overlapping with a region is translated to a query finding a set of the points overlapping with the region. However, this naive method has critical flaws. First, some queries do not produce the correct result. When the query window lies between two adjacent report time points, the query produces no result even though the trajectory is very likely to overlap with the query window. Also, it is difficult to make up an efficient nearest neighbor search algorithm because a trajectory segment connecting two location reports may be close to a given query point, but two points may be far from the point.

4.2   MBR-Based Trajectory Decomposition

Another common way to handle the monotonically growing MBR of a trajectory is to decompose the trajectory into multiple pieces. For example, when the MBR of a trajectory grows over a certain limit, decompose it into two MBRs of roughly equal volume, or simply create another MBR for the new timestamped location update. Despite the overhead of keeping multiple index entries for a trajectory, one per each decomposed trajectory piece, this approach enables to use the existing work on the spatial query processing techniques based on the decomposed MBR [23].

4.3   Time-Sliced Trajectory Decomposition

The MBR-based decomposition does not understand the semantics of time. An alternative to this approach is to partition the 3D spatio-temporal space by time slices. These time slices may be of fixed or variable length. Figure 7 shows an exemplary three-point trajectory in the 3D space decomposed by a fixed-length time slice Δt. The volume enclosed by the broken lines indicates the MBR of this trajectory, while three volumes enclosed by the bold lines indicate the MBRs of the decomposed trajectory segments. Note that the sum of the decomposed MBR volumes is a lot less than the volume of the whole trajectory MBR.

The time-sliced trajectory decomposition maintains a separate, partitioned 3D index for each time slice. When a trajectory crosses the boundary of two adjacent time slices, an implicit data point is computed by interpolation. Figure 7 shows such interpolated points by shaded dots. The MBR in a specific time slice is computed by taking these interpolated points into account. The time-sliced decomposition approach is a good match with the main-memory database

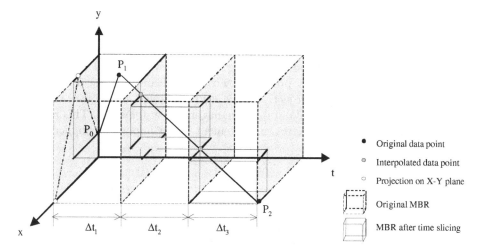

Fig. 7. A Trajectory and Time-Sliced Decomposition

implementation of the trajectory database, because the old, inactive time slices of the trajectory index may be moved to the disk-based database to free the main memory storage space.

Unlike the MBR-decomposed approach, the time-sliced trajectory decomposition has to resolve one technical issue. A trajectory starting from the time slice Δt_1 to the time slice Δt_n may not be represented in all the time slices interleaved between them because of no location update reports in certain time slices. For example, Figure 7 shows that the given three-point trajectory is not represented in the time slice Δt_2. However it shows the decomposed MBR based on the interpolation of the points in the previous and next time slices. The question is whether to store this type of implicit MBRs explicitly or not. If we choose the explicit storage, the query processing algorithms can be made simple, but it may introduce many dummy entries in the partitioned index. The trajectory of every mobile terminal has at least one entry in each partitioned index. If we choose to avoid the explicit storage, we need the run-time interpolation of the trajectory.

The time-sliced decomposition can be combined with the MBR-based decomposition in such a way that the MBR for a trajectory piece in a certain time slice is further broken down to multiple pieces. This combination further reduces the total volume of the MBRs enclosing a given trajectory at the cost of increasing the number of entries in the partitioned index.

5   Related Work

Despite much research on the main-memory database management and recent commercialization of such research, we are not aware of any work on the management of dynamically changing location information based on the main-memory database architecture. It is only recently that the database research community started to investigate on the mobile object database based on the disk-based architecture focusing on queries on the future location projection [24–26]. In addition, much research is going on the spatiotemporal data models, storage structures, and query processing within the European research network CHOROCHRONOS whose objective is the study of design, implementation, and application of spatiotemporal database management systems [27]. This research network addresses the trajectory management as an important application of spatiotemporal databases.

6   Conclusion

The need for the efficient location and trajectory management in telecommunications is expected to rise fast as the wireless telecommunication systems are mandated to process the geographic coordinate information of mobile terminals. This paper discussed the requirements on the location and trajectory management in telecommunications and brought our past experience of developing a main-memory storage system called XMAS and its spatial extension XMAS-SX to the design of the main-memory location and trajectory management database. The proposed design incorporates the embedded ODMG-compliant object-oriented model, the OpenGIS geometry model, and the location and trajectory model on top of the main-memory storage system. This main-memory database approach is expected to bring a high-performance, flexible architecture for the location and trajectory management.

Our future work includes the development of detailed algorithms for the trajectory indexing and query processing and the experimental evaluation of these algorithms by prototyping based on the existing code of XMAS and XMAS-SX. Benchmarking the main-memory database with the disk-based database with large buffer in the context of value-added mobile telecommunication services is another direction that we will pursue to prove the effectiveness of the main-memory database technology. Finally, exploring the spatio-temporal semantics is listed as a research agenda.

Acknowledgement. This research is currently supported by the ERC for Advanced Control and Instrumentation and the SK Telecom of Korea.

References

1. TimeTen Performance Software, "In memory data management", 1998. Technical White Paper, http://www.timsten.com.
2. J.J. Carery, Jr. and G.L. Stuber, "Overview of radiolocation in CDMA cellular system", *IEEE Communications Magazine*, pp. 38-44, April 1998.
3. J.J. Reed, K.J. Krizman, B.D. Woerner and T.S. Rappaport, "An overview of the challenges and progress in meeting the E-911 requirement for location service", *IEEE Communications Magazine*, pp 30-37, April 1998.

4. R.-M. Wang, "A real time fleet management via GIS/GPS platform", in *Procee-dings of the 5th World Congress on Intelligent Transportation Systems*, 1998.
5. S.K. Cha, J.H. Park and B.D. Park, "Xmas: An extensible main-memory storage system", in *Proceedings of the 6th ACM International Conference on Information and Knowledge Management*, pp 356-362, 1997.
6. J.H. Park, Y.S. Kwon, K.H. Kim, S. Lee, B.D. Park and S.K. Cha, "Xmas: An ex-tensible main-memory storage system for high-performance applications", in *Pro-ceedings of ACM SIGMOD International Conference on Management of Data*, pp 578-580, 1998.
7. J.H. Park, K.H. Kim, S.K. Cha, S. Lee, M.S. Song and J. Lee, "A high-performance spatial storage system based on main-memory database architecture", in *Procee-dings of the 10th International Conference on Database and Expert Systems Ap-plications*, pp 1066-1075, 1999.
8. T. Imielinski and J.C. Navas, "GPS-based geographic addressing, routing and re-source discovery", *Communications of the ACM*, vol.42, pp 86-92, April 1999.
9. N. Roussopoulos, S. Kelly and F. Vincent, "Nearest neigbor queries", in *Proceedings of ACM SIGMOD International Conference on Management of Data*, pp. 71-79, 1995.
10. T. Seidl and H.-P. Kriegel, "Optimal multi-step k-nearest neighbor search", in *Proceedings of ACM SIGMOD International Conference on Management of Data*, pp 154-165, 1998.
11. OpenGIS Consortium, Inc. *OpenGIS Simple Features Specification for CORBA, Revision 1.0*, 1998.
12. OpenGIS Consortium, Inc. *OpenGIS Simple Features for SQL, Revision 1.0*, 1998.
13. M.J. Egenhofer, "Reasoning about binary topological relations", in *Proceedings of the 2nd International Symposium on Spatial Databases*, pp 143-160, 1991.
14. T.J. Lehman and M.J. Carey, "A study of index structures for main memory da-tabase management systems", in *Proceedings of the 12th International Conference on Very Large Data Bases*, pp 294-303, 1986.
15. A. Guttman, "R-trees: A dynamic index structure for spatial searching", in *Pro-ceedings of ACM SIGMOD International Conference on Management of Data*, pp 47-57, 1984.
16. N. Beckmann, H.-P. Kriegel, R. Schneider and B. Seeger, "The R*-tree: An effi-cient and robust access method for points and rectangles", in *proceedings of ACM SIGMOD International Conference on Management of Data*, pp 322-331, 1990.
17. P.A. Bernstein, V. Hardzilacos and N. Good, *Concurrency Control and Recovery in Database Systems*, Addison Wesley, 1987.
18. R.B. Hagmann, "A crash recovery scheme for a memory-resident database system", *IEEE Transactions on Computers*, vol. 35, pp. 289-293, September 1986.
19. C. Mohan, D.J. Haderle, B.G. Lindsay, H. Pirahesh and P. Schwarz, "ARIES: A transaction recovery method supporting fine-granularity locking and partial rollb-ack using write-ahead logging", *ACM Transactions on Database Systems*, vol. 17, pp. 94-162, March 1992.
20. S.K. Cha, K.H. Kim, C.B. Song, J.K. Kim and Y.S. Kwon, "A middleware architec-ture for transparent access to multiple spatial object databases", in *Interoperating Geographic Information Systems*, ch. 22, pp. 267-282, kluwer Academic Publishers, 1999.
21. ITU, *ITU-T Recommendation X.722, Structure of Management Information: Gui-delines for the Definition of Managed Objects*, 1992.
22. J. Siegel, *CORBA: Fundamentals and Programming*. John Wiley & Sons, 1996.

23. Y.-L. Lee, H.-H. Park, N.-H. Hong and W.-W. Chung, "Spatial query processing using object decomposition method", in *Proceedings of the 5th ACM International Conference on Information and Knowledge Management*, pp. 53-61, 1996.

24. A. Sistla, O. Wolfson, S. Chamberlain and S. Dao, "Modeling and querying moving objects", in *Proceedings of IEEE International Conference on Data Engineering*, pp. 422-432, 1997.

25. O. Wolfson, S. Chamberlain, S. Dao, L. Jiang and G. Mendez, "Cost and imprecision in modeling the position of moving objects", in *Proceedings of IEEE International Conference on Data Engineering*, pp. 588-596, 1998.

26. G. Kollios, D. Gunopulos and V.J. Tsotras, "On indexing mobile objects", in *Proceedings of the 18th ACM SIGACT-SIGMOD-SIGART Symposium on Principles of Database Systems*, pp. 261-272, 1999.

27. "CHOROCHRONOS: A research network for spatiotemporal database systems." http://www.dbnet.ece.ntua.gr/ choros/.

Panel Session: Do the DBMS SW Vendors Offer the Products Required by the Industrial User in the Communication Industry?

Jim Gray[1] and Svein-Olaf Hvasshovd[2]

[1] Microsoft Research, California
Gray@Microsoft.com
[2] ClustRa AS, Trondheim
svein-olaf.hvasshovd@clustra.com

1 Introduction and Statements Bij Panelists

1.1 Tore Sæter; Organizer:

I am very pleased that we have a panel today with a lot of very excellent people and Jim Gray to moderate the discussion. We have split the panel in two: Three from industrial users: Ericsson, Nortel, and Telcordia and three vendors: TimesTen, ClustRa and Oracle.

1.2 Jim Gray; Moderator:

Michael Ronström from Ericsson is going to talk first. He spoke earlier today. My job is to make sure nobody goes over the 5 minutes per presentation. There will of course be questions in the end.

1.3 Michael Ronström; Ericsson:

I will speak a little about integrating databases for the telephone management network and the customer management network and databases in the layers between those. Today the network elements in the telephone management network have the original data and the customer management network have copies. What we are looking at is to try to see what can we do with having a database where we actually integrate those two things together, the real-time network approach to data and the business approach to data. We also have something in between, which serves both of those. We see a number of applications that would be interesting in this area. The data management for HLRs, which I told about in my

W. Jonker (Ed.): Databases in Telecommunications, LNCS 1819, pp. 191–206, 2000.
© Springer-Verlag Berlin Heidelberg 2000

presentation earlier today, is one such application. We also have data management for normal switches. Of course, in the new telecom networks we have data management of the Internet services, and so forth.

What would such an integrated database require? Almost everything is required to be real-time. We need real-time transaction handling. In addition to what I discussed previously we also need real-time data replication in large networks. Even if we have data replication within a cluster, we have to replicate things between different network entities. Of course, that has to be done with some kind of consistency level. It will not be based on normal two-phase commit. You have to come up with something else.

You also have to ensure that you get syntax and semantic checking, so actually the application does not corrupt the data. Of course, we also need real-time data backup. The database must always be available to business processes and network elements.

The driving force for this is that previously you had push-button telephones. The only interface you had available to change your telecom services was the push- button telephones. The wide spreading of PCs to every home means that almost every person today has a PC that can manipulate the telecom services. That means that the management interfaces on the switches will be overloaded. So that is the reason why we want to go for this new solution of integrating databases for those things.

1.4   Knut Ashaky; Nortel Networks

Where is the data? The data is everywhere! In fact, the entire network management may be built as a large distributed database, and everything that we do for information system design is applicable. In addition to that we also have to be more deterministic, and we have to provide performance. What I mean by performance here is transactions per minute.

I am going through the on-switch data management, and I am then going to move to my requirements.

This is a typical on-switch. Conceptually speaking we have the management plane and we have the switch software and hardware resources. Then we have the protocol interfaces such as a SNMP as well as the command line interface. It is important that we have a transactional interface to everything. One can look at it as a controlling environment and controlled environment. When we ask for data, some of it might be in a particular data management system on a switch. However, for some of the data we may have to go to the real world to get to it. As a result of this, the meaning of a transaction needs to be extended to the real world.

Just a quick little comment. Just like for any other information, we have to model the information. What is the information on the switch? Most people make the mistake - they stop organizing things. They organize information the way they see it in the standards. This lack of throughout organization of the information is why a lot of problems come into the products.

Again everything is just like a normal database. There is going to be transactions with ACID properties. Some of the properties might be a bit weaker, but they all got to be there. Coordinating multiple requests coming through the switch also needs to be handled. I am treating the entire switch as a database, conceptually speaking.

We are now going to go through the kind of database technology that is out there. On this stage we cannot see one size or one type of database that can fit all. I am just going to go through in a sort of broad way what technology is out there today.

We have the server databases, which is suitable for most of the applications. These server databases are the standard products we see today. Then we have the small foot print databases. They are on the mobile handhold devices. Some of them run on real-time operating systems.

Then we have the main memory databases, as well. They are very important for time critical applications. Then we have the directories that were discussed this morning. They are integral part of any IP network services of today, whether we like it or not.

Let us then move on to database feature requirements: We need to be able to have some kind of reference data type which does not exist today. On-switch we need some kind of data type that is not a value of the attribute but it tells where to find the value of the attribute. For reasons of performance we might, for example, need to limit the maximal number of instances of a table. Then I can preallocate the tables and then avoid allocation on demand. Support for both transient and persistent tables and indexes are very important. No database system supports this today.

– Support of indexes: Both main memory and disc based tables and indexes are very important. These features are very important for configurability and real-time applications.
  We would like to have checkpoint under application control. We would like to be able to say that we don't want, for example, indexes to be checkpointed but just the data. We also want to be able to tell when and where you want a checkpoint.
– Standard interface: We want to be able to use both blocking and unblocking kind of interfaces. So I can say, if you don't come back within 5 milliseconds, I am not interested or just let me know.
– Support again: Independent database file format. We want file compatibility across database products. So that I can load my entire configuration off-line and I can download that to a switch, and I can open the database right away. I don't have to do the import or export of data.

Of course, we want replication to multiple sites, and mirroring again to multiple discs.

I want to be able to say that I run the system in single user mode, so all the overhead associated with locking are avoided.

I think the point is that you also want a real-time operating system support in some cases. We also need the database running on trusted computing bases. This we see from Internet and e-commerce.

1.5   Phillys Huster; Telcordia, Director of Business Development:

I have been interested in unifying messaging for four or five years. It is my vision of unified messaging that it is more than fax and voice mail. I believe it is really a unified communication store that is going to basically push IP based content over these large mobile networks. Whether you talk about mobile networks or wireline networks, it is really indifferent to the unifying messaging server.

So what we have been doing in our labs is trying to figure out an example killer application. There is a company that came to us and said, we want to build a virtual wireless operator. This operator sits in the application space and offers back to other operators unified messaging, web browsing, WAP-services, and SMS. If operators don't have the time or money to put the services into their network, we'll just sell them a package. This involves several challenges from a database prospective. While you may have LDAP-servers and IMAP4, you only have a message store that you can manipulate. They are all separate products right now, and there is no one underlying database that can do the performance. So we are now looking at how to get a distributed real-time database that works over an IP network - that can both track and manage the activity of the customers through the unified messaging. If subscribers are roaming with their cell phone, downloading web pages, and checking their voice mails and all that, we want to track their activity. This may happen while the subscriber is moving between several operators. That means that the operators are possibly using different IMAP stores. The closest we have come is looking at things like the ISACORE meta connect product. That is a pretty cool product, but you still need an underlying meta profile, basically tracking based on a customer. That becomes a completely different issue all together.

One of the things that we would need from database vendors is a sort of "fill the gap". I give you an example: We need to media stream a voice mail file over the network extremely quickly, and we need the message store to manage that activity. So things like MP3-pushing and that sort of thing have tended to be very specialized services. We need a database that manages these kinds of activities.

Billing and service provisioning - I give you an example: We need real-time service provisioning. Someone got a cell phone that only supports certain functionality. They try to download something bigger than the cell phone is capable of handling. In his case we need to either download some software that enables the cell phone to play the file, or we need to tell the customer to upgrade his cell phone. But either way, the goal is to help the operators to get the value of its services and enhance the customers as quickly as possible without having to upgrade their backend servers.

So anyway, those are the things that interest us. We are curious to see how the vendors will approach the problem.

1.6   Omaha Shar; Oracle:

Oracle is not just a database management system company. It also has application suites and professional services. We will here focus on database management system aspects.

We provide data management services from desktop to data center that scales up, then from the desktop to chip set that scales down.

If we look at the Oracle 8i we have down to personal Oracle that run on PCs all the way up to enterprise Oracle parallel servers that run on clusters including departmental servers that run on SMP machines.

These are the range of products that we have on the high end market: We have the high end databases. Their features are scalability, high availability, manageability, and performance. We are not actually talking special data performance. We are talking about high performances for all the operations possible, and we do support ACID properties for transactions. I am not going to go into more detail here.

On the low-end side we have all the way from the chip set. We have the Oracle Lite database. Oracle Lite is two code bases that are designed for mobile database applications. They are only started to be evaluated for Telecom industry low-end applications like on-switch databases. Oracle Lite supports ODBC and JDBC interfaces as well as a more efficient proprietary interface. The Oracle Lite object kernel footprint is anywhere from 50 KB to 350 KB. The 50 KB version runs on PDAs. The 350 KB version runs on slightly larger machines. The proprietary API is designed so you can directly access data from the database and bind it to your C or C++ data structures. We eliminate lots of the overhead of going back and picking up one value at a time.

Oracle Lite has several administration interfaces. It has an optimizer. The optimizer is sort of cost based. It knows indexes and it knows the distribution of values. The indexes are all automatically maintained. It supports data centralization or replication of Oracle 8i.

In Oracle Lite you can link in your application with the database, so that the application and the database are in the same address space. It does very little copying. It maintains one copy that is used by all applications. This cuts down on processing overhead. You can have the entire database in cache. Object references in Oracle Lite are the combination of page number and slot number. It is therefore very efficient to follow these references. If you tune your database by caching it then your access time is very fast.

We allow also further tuning. You don't have to checkpoint at any time. If you want to prevent the database from writing to disc and doing an FSYNC, you can prevent that for up to as long as you want. In other words, if you can afford to loose some data or some committed changes to gain performance, we allow you to do that in Oracle Lite.

The Oracle Panama server is a server for e-business. The Panama server actually uses the Oracle Application Server and the Oracle 8i in the middle. It allows you to access any existing HTML pages and package them into transactions. It then allows an operator to provide these transaction services and the user to customize those services further.

1.7   Marie-Anne Neimat; TimesTen:

The TimesTen company is about 2 years old. Our product is an in memory database product. It is an SQL system with persistence and other features that I will talk about.

We are a high performance database company. This is the main driver behind the system. We are focused on horizontal data network.

I will now present the features of TimesTen. The system runs on standard hardware and standard operating systems. We run on HP-UX, on Windows NT, on Solaris, on AIX, and on LynxOS. We have both a 32-bit version and a 64-bit version of the system. So you can have very large databases in memory. We have databases that are up to 10 GigaBytes, all configured in main memory. The language is SQL to access the system. It is accessible through two different APIs. One is ODBC. The other is JDBC. We run an option where the TimesTen database is linked with the applications. In this option there is no overhead from context switches. We also have a client-server option if that is desirable.

Here you see TimesTen linked with the application. The database is in main memory. It is backed to disc. We do checkpointing. We have logging. We have options on when the log should actually be flushed to disc. So you can choose to have the log pushed to disc at leisure.

We are also coming out with a product this summer where TimesTen is acting as a cache for an Oracle database and is synchronized with Oracle. You can cache a subset of the Oracle database in TimesTen. TimesTen do your real-time transactions in main memory and push the updates to Oracle.

Being a standard database is extremely important. That is what allows telecom vendors to come up with new services in a reasonable amount of time and be competitive on the market. This makes it easy to administrate services. The performance is what we are the most proud of. The performance competes with home grown data managers. At the same time it gives you an off-the-shelf product with standard APIs.

The system scales up to tens of thousands transactions per second. In other words it scales up to millions of subscribers. We have the high availability option through replication. We do support conflict resolution. The system is embeddable. You can link it with the application. The footprint is fairly small.

Just one slide on performance: This is a subset of the Wisconsin benchmark. Some of the top queries are single table selects. The next ones are multi-table joins. The system is being compared here to a disc-based database where the database is entirely loaded in main memory. So we are comparing the two systems apple-to-apple. The results show a factor of ten in speed-up for TimesTen.

1.8   Svein-Olaf Hvasshovd; ClustRa:

Like Marie-Anne I will tell you a couple of words about the company first. The company is 2 years old, it is a spun off from Telenor, the main Norwegian telecom operator. We are now an American company with headquarters on Route 128, Boston, MA. The company is specialized in mission critical telecom databases. You can see here what kind of main technical features, we have put into the system: Continuos availability, real-time response, linear scalability, off- the-shelf based, and SQL interface.

I have looked at a lot of other peoples slides today and found that real-time are at least repeated three times at the top of everybody's requirements list. We think that mission critical telecom databases really demands extremely high availability. When somebody talks about 30 seconds down-time per year, we are taking that seriously and make our product meet that kind of requirement. This level of availability not only requires fail-over but also automatic repair, dealing with online scaling of the system, dealing with physically moving of the system while it is fully available, dealing with online upgrade of software, dealing with all kinds of software and hardware failures without disrupting transaction services. All this is done on fully available system serving thousands of transactions per second. The only thing we do not currently support is complex online schema modifications.

We combine extremely high availability with real-time. You see 15 milliseconds response times here on the slide. The slide is old. We currently go below 1 millisecond on TPC/B-like transactions. ClustRa runs off commodity hardware and basic software. We agree with everybody else in the industry on standard interfaces. We provide ODBC/SQL and JDBC/SQL. In addition we provide a lower level real- time interface.

If you compare this list of bullets here with the earlier requirement list, for example, Ericsson came up with for mission critical telco databases, you will find that there is a big match in what we provide, and the telco industry demands.

2   Discussion

Jim Gray, moderator, opened for questions from the audience.

2.1   Question:

I am trying to reconcile what I am hearing at this panel with the opening keynote talk. The opening keynote talk was saying that databases seam to have lots of success stories in other applications but in the telecom business the situation seams to be somewhat more mixed. From what we are hearing in the panel it sounds somewhat more honkey-dory. All we need is more of the same.

Michael Ronström; Ericsson: Well at Ericsson we have a long history in doing everything ourselves.

2.2   Question:

Would you say all telecom business do that?

Michael Ronström; Ericsson: I can only talk for Ericsson, but I guess it is the same for other equipment manufacturers. We have, of course, done analyses of open database systems. Most of these components come from pretty new companies. There is a shift on-going in Ericsson. There are more requirements for open software.

Marie-Anne Neimat; TimesTen: We see more willingless for open software. One of the reasons for this is that service providers have to deliver services faster and to do this they can not do everything themselves in-house any longer.

Svein-Olaf Hvasshovd; ClustRa: The development cycles in the telecom industry have switched from five to ten years to develop a new product down to currently more like 1 year from the start of a project to when a new product has to come out. As a result telecom suppliers don't have the time anymore to develop everything themselves in-house.

I have an earlier background from telco research. As a researcher I was allowed to investigate a number of switch products developed in the late 70s and early 80s. I found to my amazement that inside these products there were advanced databases embedded. These databases were especially advanced with respect to availability. None of this early database development done in telco DBMSs has ever been published through academic channels and is therefore almost unheard of outside the teams that built the products. There have been a number of very successful databases in older telco products - but we have not heard of them.

Knut Ashaky; Nortel: I agree with Michael Ronström (Ericsson) in what he said. We used also to develop everything ourselves in-house from compilers, and databases to communication protocols. The whole thing was developed in-house. What we see now in the industry is that lots of companies start buying small start-ups or they start buying off-the-shelf products, to buy time-to-market. We can't be for everything anymore. We have to focus on things we are really good at which are telecom itself. We don't have to do databases - let the database companies do that, and we become the users of them.

I would like to add that in-memory databases have only been available for 3 to 5 years. It is now half a dozen companies working on this out there. Main memory databases don't yet have all the kind of features we are looking for. But certainly it is much better than it used to be. I think that it a good sign. Things are getting better in that respect. They are much better than they were 10 years ago.

2.3   Question:

As I listened to the talks today I started looking at all the application domains people were listing here as telecom databases. They include billing, financial, e-business, and so and so. I wonder if you can tell me what is not a telecom database? If we cannot define it, why do we have this special workshop on telecom databases?

Knut Ashaky; Nortel: Telecom applications are filling the same area, in some aspect, as the ones that you mentioned. They are just some sort of traditional databases. But you will also find special telco databases like you find special databases in process control and air traffic control. Those kinds of areas are also into very specialized databases, because of the performance. They have been using specialized databases for a long time. The specialized databases have always been there but not been published on.

2.4   Question:

But the specialized databases are not telecom databases ...

Knut Ashaky; Nortel: No, they are not. But again, what is a telecom database. As I have shown in my picture, data is everywhere. What is telecom specific data? It touches everything: Data counting, data mining, HLR, VLR kind of databases. So it is all over the place. Data is everywhere. So it is some specific part of it. There are some specific features, but other domains require them too.

Michael Ronström; Ericsson: In my view telecom databases are databases involved where the users are directly involve in the usage of it. Actually billing databases would not fit into that category. So I would say, billing databases are databases used by telecom companies. Real-time billing, for example, would fit into the telecom database area. That is how I would define a telecom database.

Jim Gray; moderator: Just parenthetically. In the US the telecom companies are becoming Internet service providers. They are hosting web sites. They are offering online backup. They are giving people online billing and online banking. They are going to every business that they can go into legally.

2.5   Question:

I am repeating, what is not a telecom database?

Marie-Anne Neimat; TimesTen: I think that the gentleman from Ericsson said it well. There is a consumer sitting there as part of the transaction, and data and the transaction have to be delivered in a reasonable time to the consumer. If there is not a consumer in the picture then it is not a telecom database.

Omaha Shar; Oracle: But on the other hand if you look at the requirements. They are the requirements for database management systems all along. Like fast response time, scalability, and availability. So essentially the only new thing here is the real-time response for some telecom applications.

Svein-Olaf Hvasshovd; ClustRa: My answer to the question is, there exists no telecom database as such. What you find is that telecom tends to take a number of database requirements to the extreme. You find extremely requirements for real-time. You find extreme requirements for availability. You find extreme data volume requirements. So my answer to the question is, telecom tends to take requirements up to the extreme, and in doing so they are sitting on the edge of database technology.

Willem Jonker; Organizer: I would like to add a few words. The title is not telecom databases but Databases in Telecom. That is very important. There is not such a thing as a telecom database. I think what is important about the thing is that we look at the specific telecom applications. They come up with specific requirements that may come from other areas as well. Their demands may be covered with technology that originates from other things. I wouldn't care and I agree with what Svein- Olaf is saying. I think what is specific about telecom applications is that they drive things to the extreme. Scalability, its large size - in terms of space - in terms of response time - in terms of number of users. Which bank has 2 million users online. So it is driving things to the extreme. I think that is the main thing. I don't care which database technology will solve it. If a single Oracle database will solve all the telecommunication problems, I am happy with that. But, I don't think it does. We should not try to build some database that is a telecom database. We should look at telecommunication applications. See what kind of extreme requirements are there and which really counts there and open new ways of managing data. For me it is also important to think even broader than a database management system. I would say it is more like data management in a telecommunication network. As was said in the panel before, there is data everywhere, and most of the data in telecom communication systems are not even in databases.

2.6 Question:

Another important issue is integrating and combining information from all the sources. We want to get an integral picture of our customers. To do this we have to know customer data, traffic data, and so forth. All this data is coming from different sources; from switches, from flat files, from databases. We have to integrate. We have to know our customers better.

2.7 Question:

What we actually need is information bases. Look at the market capitalization of any telco. Only about 25market capitalization is actually accounted for by network, physical access, the contraption. 4/5th of the market capitalization in telco is in intangibles. That is the really asset of the telco's: information! It is the lack of meta data we are interesting in which is hardly mentioned today. It is a sort of appalling management of meta data in telco industry. It is the really issue. I would like the panel to comment on whether there is any hope for better management, better provision of meta data which actually is needed to give information.

Omaha Shar; Oracle: Any model of meta data are using the database management system. You could certainly create a schema to model your meta data.

2.8   Question:

Oh yes, there are plenty of good information models around. These are also some very complex meta data schemes. The object management group made a great progress on that. The technology is there. The lack of meta data management is a culture issue. Why are not telco companies better managing and producing the meta data?

Phillys Huster; Telcordia: That is an easy one to answer. Because telco has traditionally been regulated by the government they are not competitive creatures and as result meta data have not been of significant importance for them. Telcos are just now becoming competitive creatures. As a result we will see a shift in attitude towards meta data. A place to look for good meta data management will be the ISP market. Those guys are doing brilliantly in terms of the meta data. Look at Amazon. They have a package they used to call Broadvision that does a phenomenal AI customer tracking building of knowledge. Meta data is a keystone to the ISP market. Then you have the concept of the info mediator, which is these third parties that are going to buffer you from a trust perspective to the ISP's.

So the real challenge here is a race to the services. It is not a race to the lower layers. The wireless operators have enough stuff in place to do some cool things. It is really simple, but how many wireless operators are really doing fax on the cell phone?

So I agree with you about the entire world is really going towards this meta data part. I like to see more research in these areas.

2.9   Question:

I think you touch on intelligent concept based searching. The idea is to sort out this information from the more simplistic approach like the very simple catalogues that are available.

Phillys Huster; Telcordia: The root goes back to expert systems that nobody liked but neural network and expert system have crossbred nicely. It is just where have they ever impacted database community in terms of the conceptual models. That was my frustration even back in college.

Knut Ashaky; Nortel: I have another slide! Actually I didn't sort of go through the whole thing.

All the configuration of the application data must be accessible from system catalogue objects of the database. I think you have to manage the meta data. In other words, distribute the meta data. Otherwise, you are not managing the data. If you are looking for extensibility you have to capture the meta data. For example, on our switches we don't want a sort of recompile everything as soon as something changes. That distributed meta data must be part of the entire management of the data and be accessible just like any other application data. All I want to do is just make that change in the meta data and the application data will be affected automatically.

2.10   Question (Jim Gray):

In the United States common carriers have to report their availability. Every time they have an outage affecting more than 50,000 people, they have to report the outage. They deliver 3 to 4 nines (Class 3 to Class 4) of availability. The outages are actually pretty bad. The biggest problem is overloads.

There are a lot of telecom people here. I don't know if this is a worldwide thing or just an American thing. There is fundamentally some lying going on here, which is that the telecoms are claiming they are delivering 5 nines (Class 5). MCI took their frame relay network down for a weekend. I don't know how they accounted for that. In general, the reports they file show that they deliver 3 point something nines. This is at all not that different from what the Schwab (on-line broker) or emails deliver. I don't know if other people in the panel want to comment, but there are two things, I heard: 30 seconds of outage per year as the requirement, and 3 millisecond response time. When I make a phone call I am astonished when I get connected within a second. So I am puzzled.

Michael Ronström; Ericsson: Come to Sweden!!! In Sweden you are connected within a few hundred milliseconds. I mean, a user doesn't know this well, whether the connection time is less than a second or not.

Phillys Huster; Telcordia: This goes back to either having a deregulated highly competitive and therefore not a government regulated industry or you have a governmental regulated one. In the case of MCI going down - the same thing happened to AOL - they could go down for a day or two days without loosing a single customer. The point is, it is whatever the market demands at this point the telco market delivers.

Michael Ronström; Ericsson: I can mention a few things about what I know from Ericsson. Of course, I cannot say a thing about the exact figures because that it is confidential, but what I have seen is in some markets we actually meet the requirements but it is very much depending on the operators.

Phillys Huster; Telcordia: There need to be these universal government parties that encourages everyone that follow certain specifications and there should be incentives for the vendors to pay the money to get up to that level of quality.

2.11   Question:

The 5 nines actually have some penalty with it which is basically when you sell an equipment and you sell a 5 nine. If it goes below this, you actually start paying back. This costs you dearly. When AOL goes down - it comes to "well we give you a month free" - 20 dollars are nothing. They can afford doing this.

2.12   Question:

Do we have the same problem with the scalability? We have in Netherlands scalability problems with prepaid cards. The switches became overloaded because the prepaid cards were so popular that the switch could not cope with it and the database access could not be made. So what was done was that the database access were dropped so most of every three calls out of ten were free without people knowing it. Finally they did find out. The company was actually fined by the authorities and we were forced to pay back money to the people that have bought prepaid cards. So I think when telecommunications go into these areas where they found more and more to do with database information technology which are less stable than let's say the good old switches that they used to, that they have to find a way to deal with this, because this can lead to bankruptcy if the authorities are getting really strong on that. It also affects the competition on card product. Our competitors were able to provide the service because they did not sell that many prepaid cards. It didn't run into the same scalability problems in that form. I think I see degrading quality service within telecoms when service complexity grows.

Svein-Olaf Hvasshovd; ClustRa: It is hard to answer you, Jim, on the relevance of your experience in US to Europe. So let me leave that aspect out of my answer.

The thing I want to comment a little bit on is the number of nines. I have experienced that the number of nines is strongly correlated to the revenue of the service. When the revenue is low which correlates to the number of subscribers for the service being low - the number of nines is low. As the revenue increases, the availability also increases. The traditional fixed line phone services which still represent the highest revenue have typically a higher availability than mobile services which again have a higher availability than intelligent network services which represent rather low revenue. At the same time, I have experienced that the requirements for nines for mobile equipment is fast growing as the mobile revenue becomes comparable with fixed line revenue.

Concerning the degradation of service, I see both a trend of increasing service availability and lowering it. For some highly sensitive services where phone, Internet, and TV channels are bundled together even more than 5 nines are now required. The same trend is found in mobile services - increased number of nines. On the other hand in new services fewer nines are required.

2.13 Question:

I could mention a few things on response time. We have discussed the response time of a few milliseconds. The response time when you are lifting up the phone also include for example the number 7 network. When we did the simulation in the UMTS-project, the set up time requirement was 0,6 seconds for the entire network. That boiled down to that the database part of the project was 5 to 10 milliseconds.

2.14 Question:

I have seen cases where different services have been sold to different kind of customers on different availability level like a virtual private network service. If the carrier wants to sell that service to the police or the armed forces, they can sell it at a higher price but then have to guarantee a higher availability level. If they want to sell to, for example, a fast food chain, they can probably sell it at a lower price but at a lower availability. So flexibility in availability is probably going to be something which is very attractive to the carriers.

Phillys Huster; Telcordia: I can pick up my phone and make a phone call, but I cannot send a file with a telephone call. One reason for this limitation is the unwieldy SS7 network. I think there is an opportunity for us to improve the quality of the communications in these new IT based networks.

There are people like me. I use email constantly. Servers go down and the email doesn't get there. This is good enough for me as an ordinary user. Will I pay for a quality email service? Absolutely, I would love to know that every email I send is going to get to the person I send it to. So I think there is an opportunity for the capitalistic market. In the older days it was toll grade or nothing and now there are going to be gradation of service.

2.15 Question:

One definition we did earlier in the discussion for databases in telecommunication was whether they demand extreme performance in some respect. What is extreme today is very often easy tomorrow. So therefore the question is, what do people think will be the characteristics of the key databases in communication in 2 and 5 years time?

Michael Ronstöm; Ericsson: What we have seen at Ericsson is that when the processors get faster, customers require more functions. So the matter is who is moving the faster, the processor development or the customer requirements.

Jim Gray; Moderator: I had assumed that the panel would say, that when you use commodity hardware, then you have huge savings and therefore you can over-provision and you can use high-level programming languages instead of assemblers and so on. I think we are still in that space.

One of the things I disagreed with in the panel is, they said that telecommunication is fundamentally directly involved with people. I think five years from now that it will mostly be computers talking to computers, and it is going to be the thermometers talking to furnaces. It is going to be meters talking to billing machines.

2.16   Question:

Would they be talking SQL?

Jim Gray; Moderator: Maybe, there will be some databases in the middle there recording things.

My premises are that: In the beginning it was punch cards talking to computers, and then it was dumb terminals talking to computers, then PCs talking to computers. Then the front office got automated. Now the customers are getting automated. Pretty soon the customers are going to have robots that are talking to each other. So that is the progression that we see, and incidentally it is going to be completely automatic - completely brain dead, up all the time, and self-healing. That is the requirements but few people are talking about these self-managing systems.

PANACEA: A System That Uses Database Technology to Manage Networks

S. Seshadri and Avi Silberschatz

Bell Laboratories, Murray Hill, NJ 07974
{seshadri,avi}@research.bell-labs.com

Abstract. Automated tools are crucial for managing the emerging large-scale complex and heterogeneous networks to ensure that networks remain healthy and available. In this paper, we first point out deficiencies in current network management tools and methodologies and then advocate the use of database technology to ameliorate these deficiencies and propose a generic architecture for network management on large-scale networks. We also identify a number of of database research problems that need to be solved before database technology can be successfully integrated into network management.

1 Introduction

Due to the explosion in the complexity of the networks in the last decade, network management has become critical. Network management is required to perform fault diagnosis, performance management, predict loads and plan for future traffic, etc. Automated tools for network management on such large-scale complex and heterogeneous networks are crucial to ensure that networks remain healthy and available.

Automated tools for network management can be characterized, by and large, into three stages:

Data Acquisition: This stage involves collecting data about the activity of network elements such as routers, switches, managed hubs etc.
Data Analysis: In this stage, the collected data is analyzed for various indicators of faults or performance problems such as the fraction of packets that have checksum errors on a particular interface of a router.
Operations Decisions: In response to the analysis, operations decisions, such as, change the routing table entries to bypass a faulty link, are taken.

In this paper, we first point out deficiencies in current network management tools and methodologies. Later, we advocate the use of database technology to ameliorate these deficiencies and propose a generic architecture for network management on large-scale networks. We also identify a number of of database research problems that need to be solved before database technology can be successfully integrated into network management.

W. Jonker (Ed.): Databases in Telecommunications, LNCS 1819, pp. 28–39, 2000.

The Simple Network Management Protocol (SNMP) is the dominant standard for network management today. In fact, it is used as an acronym for Internet-standard Network Management Framework. SNMP was proposed in the late eighties, when network sizes were small and in fact was proposed as a stop-gap measure. In this paper, for concreteness, we focus on SNMP and in particular, highlight the deficiencies of SNMP and propose solutions to ameliorate these deficiencies. However, other protocols like CMIP and the proprietary ones suffer from the same problems in addition to lack of widespread acceptance by all vendors.

SNMP and the associated network management methodologies suffer from the following deficiencies:

Generate High Volume Of Management Traffic: The SNMP protocol supports retrieval of single objects stored at the network element but does not allow any sort of computation to be performed at the network element. As a result, large volumes of data may need to be transferred to the network manager (the station where network management is being performed) and the manager may filter most of the retrieved information.

No Support For Event Notification: Although there is some primitive support for event notification in the form of traps in SNMP, it is not sufficiently expressive. Therefore, network management using SNMP is predominantly polling based, which results in the familiar problems of either missing an event (if the polling interval is high) or incurring a huge traffic overhead (if the polling interval is low). To perform effective and efficient network management, support for complex event detection and notification is required. For example, a network manager may want to be notified when the average error rate on all the interfaces of a switch exceeds ten percent.

Centralized Processing: Network management has traditionally been performed in a centralized fashion primarily to ensure that the impact of adding network management to managed nodes is minimal. However, the central network manager could become a bottleneck as the network complexity increases. In recent years, the computational power of the network elements has grown tremendously, thus making it possible to perform significant network management functions in a distributed fashion.

In this paper, we outline the design of a system called $PANACEA$[1], which addresses the above deficiencies. The basic idea behind $PANACEA$ is to enhance network elements to i) process declarative queries and ii) support triggers. This can drastically reduce the volume of network management traffic and further, distributes the intelligence and control in the system.

Additionally, $PANACEA$ also supports the notion of auxiliary network managers (ANM) which are useful in two different contexts: (i) not all network elements may support the enhancements above in the short term, in which case an ANM running on a machine close to the network element can proxy for the

---

[1] $PANACEA$ is an acronym for PlAtform for Networked Active Capability basEd dAtabases

network element thus still effectively reducing network management traffic between the ANM and the network manager and (ii) the ANMs can be used to combine results from multiple network elements (even if the network elements supported the enhancements above) that are close to it. The combined result may be much smaller than the sum of the original results.

The above architecture raises several interesting technical and research issues. Some questions that need to be answered are:

1. How exactly is the database functionality pushed to the network elements?
2. How many auxiliary network managers should we place and where should they be placed?
3. How are complex events detected in an efficient manner?
4. If data were to be stored in XML (as is the current trend), how would our system adapt to it?

The rest of the paper is structured as follows. Section 2 explores the deficiencies of SNMP based network management in detail. We describe the architecture, design and the current state of the *PANACEA* system in detail in Section 3. Section 4 identifies several interesting and novel research issues that arise in the context of *PANACEA*. We identify connections with other related work in Section 5 and conclude in Section 6.

2   Drawbacks of Current Network Management Methodologies

As we mentioned in Section 1, SNMP has emerged as the standard for network management in the internet and is used as an acronym for the entire Internet-standard Network Management Framework. SNMP has two important components:

1. The first component is the notion of a Management Information Base (MIB), which is essentially a schema for storing data objects related to a network element's activity.[2] The schema is essentially a hierarchical database in that the entire data is organized as a tree. We illustrate this using a small fragment of an address forwarding table in the Bridge MIB (RFC 1493) [1] which is depicted from a relational database viewpoint in Figure 1. However, the same table is stored in the MIB as a tree as shown in Figure 2. Each node in the MIB is identified by the concatenation of the labels of the nodes on the path from the root. For simplicity we have assumed Fdb as the root (in reality Fdb would be a subtree of a larger tree). Note that the entire column corresponding to MACAddress is stored as a subtree rooted at Fdb.MACAddress. Fdb.MACAddress.1 denotes that we are referring to the MACAddress column of the first row of the table. The value of a node (if applicable) is shown in parenthesis below the identifier for that node. For example, 12 AB EF 34 21 13, is shown in parenthesis below Fdb.MACAddress.1.

---

[2] MIBs have become so popular that standard MIBs are now defined for anything that is accessible through the network including proxy caches, databases etc.

2. The second component of SNMP is a standard protocol for retrieving information stored in the MIBs. This standard allows network management software to retrieve specific objects (using snmpget) in the MIB or retrieve an entire subtree (using snmpwalk) rooted at a node. For example, "snmpget Fdb.MACAddress.1" would return 12 AB EF 34 21 13 while "snmpwalk Fdb.MACAddress" would return the entire column corresponding to MACAddress. Similarly, "snmpwalk Fdb" would return the entire table. In addition, SNMP also allows individual variables in the MIB to be updated.

We will now illustrate in detail the inadequacies one by one in the following sections.

2.1   Lack of Support for Simple Queries

The first and the most important deficiency is lack of support for simple queries. This results in a huge amount of network management traffic since all the raw data required for answering the query needs to be shipped to the network manager rather than just the result of the query.

MACAddress	PortNumber	PortState
12 AB EF 34 21 13	1	Forwarding
12 AB EF 34 22 13	1	Blocked
12 AB EF 34 21 14	2	Blocked
12 AB EF 34 21 23	3	Learning
.		
.		
.		
.		

Fig. 1. Address Forwarding Table

Suppose the network manager is interested in the answer to the query "Retrieve the MAC addresses associated with ports that are in the forwarding state" on the table of Figure 1. Using SNMP, there are two options to retrieve the desired information:

1. The first option is to retrieve the entire table and filter out rows for which ports are not in the forwarding state. This could however be very expensive

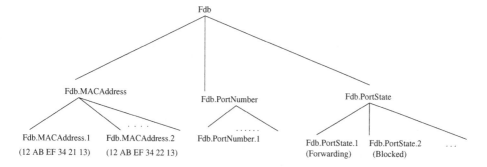

Fig. 2. MIB Structure

since the table may contain several tens of thousands of entries given the complexity of the networks and the blurring of the boundary between Layer 3 routing and Layer 2 switching functionality. Thus, we may end up retrieving several megabytes of information, although we are finally interested in a few kilobytes (there may be very few ports that are forwarding). For example, in Figure 1, only one of the four rows shown is relevant for this query. The situation gets much worse if we are only interested in some aggregate quantity.

2. The second option is to retrieve the column corresponding to the port state using "snmpwalk Fdb.portState" and then for each row for which the state is forwarding, fetch the corresponding MAC addresses. This results in one SNMP request per matching row which may mean several thousands of messages.

Further, notice that each network management software (the clients of SN MP) has to perform its own query processing. Thus, SNMP provides a very primitive and low level interface to its clients.

2.2   No Support for Event Notification

Most aspects of network management like performance management inherently rely on being notified of certain abnormal events. Thus, event notification is invaluable to network management. While SNMP has some primitive form of support for event notification in the guise of SNMP traps, these are not sufficient. SNMP traps are used to notify events like the failure of a link but not events that involve the outcome of complex computations on the MIB data. For example, a network manager may be interested in being notified when the average packets dropped due to a checksum error on all the interfaces exceeds 10%.

In the absence of rich event notification capability, current network management software resort to periodic polling. This suffers from the classic problem that if the polling interval is too small, there is a huge traffic overhead while if the polling interval is large, we may miss the event of interest or recognize it too late.

2.3 Lack of Consistency in the Returned Results

Another subtle problem that arises when SNMP is used to retrieve large volumes of volatile data is that the returned result may not be consistent. For example, when the address forwarding table of Figure 1 is being retrieved, one column of a row may be retrieved but not the other. Recall that each column of the table is stored as a subtree and therefore "snmpwalk" returns the first column of all the rows before it returns the second column of the rows. However, entire rows are deleted or inserted by the network element corresponding to a certain activity. For example, in the address forwarding table, a row may be deleted since the entry has aged[3] after the first column of the row has been retrieved by a "snmpwalk" but before the second column is retrieved.

In the example in Figure 2, we have used physical row numbers to match columns of a given row (i.e., Fdb.MACAddress.1 is matched with Fdb.PortState.1). Notice that if the matching is performed with physical row numbers, then it may be impossible to even realize that there is a consistency problem. However, in general more immutable identifiers are used at least in some tables. In this case, network management software has to interpret the returned results carefully and complex logic is required to match corresponding columns of a row.

2.4 Discussion

It should be apparent from the above discussion that current network management methodologies suffer from significant drawbacks. Many vendors have also developed proprietary interfaces to retrieve information from the network elements. These interfaces also suffer from the problems mentioned above. More importantly, the proprietary interfaces render the software for a heterogeneous multi-vendor network hopelessly complex.

To counter the above problems, clients of SNMP have to incorporate complex logic for query processing, event notification and consistent retrieval. It should be apparent that database technology can be used profitably in this environment and we discuss this in the next section.

3 *PANACEA* Architecture

We propose a system called *PANACEA*, illustrated in Figure 3, to address the problems identified in Section 2. In Figure 3, NM is the station at which network management is performed[4] while NE1, NE2, etc. are the network elements (managed entities) in the network. The NM interacts with the network elements and retrieves information from them as well as informs them about the events it wished to be notified about. The goal is to enhance every network element

---

[3] Learning bridges age their entries every so often (five minutes by default) to allow for machines/network cards to move locations)

[4] In general, there could be multiple network managers in our system but we restrict our attention here to the case where there is one, for simplicity of exposition

with database technology[5], however, this requires some support from network element manufacturers. Therefore, our architecture does not assume all network elements have been enhanced with database technology, but yet tries to minimize the impact of these network elements. In Figure 3, NE1 and NE6 are network elements that have been enhanced with database technology and are denoted by the rectangle around these network elements. This enhancement enables them, for example, to perform some computation at the network element and return only the information that the network manager is interested in thus minimizing the volume of network management traffic.

We use the notion of an Auxiliary Network Manager (ANM) to minimize the impact of network elements not enhanced with database technology. For example, NE4 and NE5 are not enhanced with database technology but ANM2 performs the desired functions on behalf of NE4 and NE5. Thus, ANM2 can minimize the volume of traffic between the NM and ANM2. This is especially crucial if the NM is connected to ANM2 by a slow WAN or wireless link. Further, even if NE4 and NE5 can perform query processing, ANM2 can combine results from NE4 and NE5 and then return a possibly smaller result to the NM.

A major concern with pushing database functionality into network elements is the extra load we may be adding to them. The explosion in processor speeds coupled with the trend towards pushing some filtering functionality, into SNMP [6] as well as proprietary interfaces, makes us optimistic that this load will be manageable. However, we want to start small in the first phase, and keep the extra load on the network elements to a minimum, and add more functionality based on our experience in the first phase.

In the first phase of the project, we propose to add the following functionality to the network element:

1. Add support for executing single block SQL queries, consisting of SELECT, FROM, and WHERE clauses only, at the network element. Thus, we are not considering nested queries and group by at this stage. In fact, additionally, we may even allow network elements to support only selects and not joins. Note that although we plan to use SQL initially, we realize that the SNMP protocol is very dominant in the network management community and can not be replaced overnight. One option is to consider extensions to SNMP, as is in fact being attempted in [6]. Alternatively, we could ensure backward compatibility with the current SNMP standard. The ultimate winner will be decided by various players in the network management community. Our goal is to keep most aspects of *PANACEA* reasonably independent of the language so that the technology developed in *PANACEA* remains useful irrespective of the outcome.
2. Add support for constraints expressed in terms of the above subset of SQL. These constraints will form our initial event notification mechanism – the network manager is notified if there is a constraint violation.

---

[5] We use this term loosely to refer to some form of query processing, event notification and support for consistent retrieval in this section. We make this more precise in the next section.

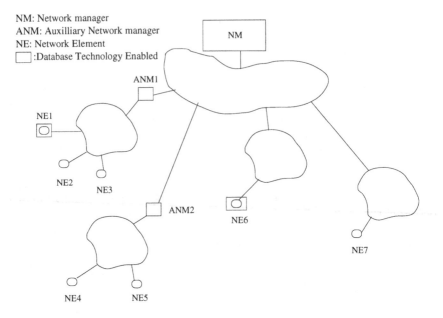

NM: Network manager
ANM: Auxilliary Network manager
NE: Network Element
☐ :Database Technology Enabled

Fig. 3. Overall Architecture of *PANACEA*

We examine the technical and research issues involved in realizing *PANACEA* in the next section.

4 Research Issues

In this section, we discuss some of the technical and research issues that arise in the context of the *PANACEA* system.

4.1 Support for Database Technology at the Network Elements

Network elements typically run proprietary operating systems with very little support for enhancing their functionality. However, of late, there is trend towards supporting a Java Virtual Machine (JVM) inside routers and switches and possibly Jini in the near future. Therefore, we plan to use Java as the development language for *PANACEA* and install what we call a Database Virtual Machine (DVM)[6] on these network elements.

An alternative often proposed, for example in [5], for enhancing the functionality of switches and routers is to supply required Java code along with the data. In our case the query is the data and the part of the query processing engine relevant for answering the query is the code. We feel this is not the right choice for two reasons:

---

[6] This term is also used by the Jaguar project at Cornell University [7]

1. First, in our case the set of functions that need to be applied on the data is a small one (selects, projects, joins, etc.). Shipping code with the data makes sense only if a) the entire code base is very large that it can not be stored at the network elements and b) figuring out which code path to take for a particular data is a complex function of the data and other parameters that we do not wish to load the network elements with this added complexity. In our case, if we were to ship a query execution plan that is understood by the database virtual machine, then we just need a simple execution engine that would execute the plan. Thus, we would not perform optimization at the network element, and minimize the load on the network element. We are currently investigating an appropriate representation for these query execution plans that are universal and can be transported to any network element.

2. Second, there is no simple way to check for constraint violations since updates are performed by the network element itself.

4.2   Management Information Bases

As we mentioned before, the MIBs are valuable and all pervasive. We, therefore, propose to continue to support the MIBs as they are today. Some of the variables in the MIBs are updated by the network element based on the activity at the network element. There are two choices for where exactly to store the MIBs and we discuss the pros and cons of these choice below:

1. The first choice is to locate the MIBs under the control of the vendor's software as is done currently. This implies that the DVM needs to be given a mechanism for accessing the MIBs to answer queries. The DVM should also notified when the MIBs are updated so that auxiliary indexes and views (to aid answering queries and detect constraint violations) at the DVM can be maintained consistent. One advantage of this choice is that the SNMP and the proprietary interfaces will continue to work uninterrupted and thus guarantee backward compatibility with SNMP.

2. The second choice is to locate the MIBs under the control of the DVM. In this case too, the DVM has to be notified whenever the MIB needs to be updated, by calling an appropriate update function. This choice solves the consistency issues mentioned before, since the complete data is under the control of the DVM. The DVM could continue to accept SNMP requests for backward compatibility. We have chosen this option since the MIB can be stored in an efficient manner by the DVM.

4.3   Auxiliary Network Managers

As we mentioned before, not all network elements will be enhanced with database technology in the near future, and we propose using Auxiliary Network Managers (ANMs) to act as a proxy for such network elements. In response to a query from the NM, the ANM retrieve the required information from the network element using SNMP. The ANM polls the network element to track the values of appropriate data elements to help it notify the NM of the events that the NM is interested in. Finally, the ANM ensures that the results returned to the NM are consistent.

Given a certain budget (number of machines that can run ANMs), we need to decide where in the network they should be placed. This depends on the speeds of the various links in the network, the traffic pattern (for network management) between the network element and the network manager etc. We propose to investigate this issue in the future.

4.4   Complex Event Detection

In the first phase, the events we are interested in are constraint violations. We propose to use auxiliary views to help detect such constraint violations efficiently. The set of auxiliary views to be maintained has to be determined in a cost based manner depending on the constraints and the rate and nature of updates. Further, the constraint definition may intrinsically have some sharing (due to common sub-expressions). We plan to investigate this issue further.

4.5   XML

In a few years, most data on the network elements may be stored in XML. Web based network management is increasingly becoming popular. In the next phase, we plan to investigate issues that arise assuming data is stored in XML. Traditional issues like query optimization, query processing, indexing and storage have to be revisited in the context of XML. Research into semi-structured data (for example see [8]) has made a beginning but there are many issues still unsolved. Further in our context, we are primarily interested in XML data that is resident in main memory.

4.6   Query Optimization

Notice that the *PANACEA* is essentially a distributed database system with each site having differing capabilities (As we mentioned before, some sites may perform only selects while others may perform joins too). Thus, there are many possibilities for executing a query which involves base relations from multiple network elements. The optimizer needs to produce a global query execution plan that is "good". A global query execution plan annotates various portions of the plan by the location where it gets executed taking into account the capabilities of the sites. Notice that to minimize network bandwidth, there may be a choice for example between moving the result of a join and the operands of a join.

The queries posed by network management tools in close succession are likely to access similar data. We could cache the final and intermediate results of queries and use the cached results to answer future queries. We need to address the issue of how we optimize an incoming query with respect to the current state of the cache. This is similar to query optimization in the presence of views, but we may need cheaper alternatives since the optimization is performed at execution time. We also need to work out mechanisms for cache management. Cache management is complicated here by the fact that cached elements are variable sized and are not independent (a cached element may be derivable from other cached elements).

5   Related Work

The most closely related work is the ongoing effort at adding some query processing capability to SNMP [6]. This is a reasonable approach in the short term, but extending SNMP beyond a certain degree would get very cumbersome. The advantages of a declarative language have been proved several times over and we believe that our approach is more appropriate in the long term.

The notion of active networks [5] in which customized programs are injected into the nodes of the network also attempt to enhance the functionality of nodes in the network. However, as we mentioned before, we believe they are not appropriate for our purposes since the degree of customization required is extremely small and further event notification is not simple with active networks.

The notion of pushing database functionality into embedded devices is not entirely new. The Jaguar project at the Cornell University [7] is also investigating this issue. Cloudscape [2] and Databahn [3] are new companies that have entered the embedded data management market. Conventional relational vendors are also entering this market, see for example, Oracle Lite [4]. Our interest is not just embedding a database virtual machine at the network element but utilizing the distributed, capability-based architecture of *PANACEA* to our advantage.

6   Conclusions

Current network management methodologies can result in excessive network management traffic due to lack of support for queries and event notification. We argue that network management can benefit from database technology and propose an architecture of a system called *PANACEA* that pushes database functionality into network elements. Our architecture takes into account (i) the difficulty of displacing existing standards and (ii) the overheads of a full-fledged database system.

PANACEA throws up several interesting and challenging problems. The actual mechanism for pushing database functionality into network elements, placement of the ANMs, complex event detection, handling XML data and query optimization are some of the issues that need to be addressed to successfully integrate database technology into network management.

We have an initial design of the first phase of the project and the implementation of the first phase is under way.

Acknowledgements. We thank Yuri Breitbart, Minos Garofalakis, Rajeev Rastogi and Aleta Ricciardi for several discussions on this topic. We also thank Zhenghao Wang for assisting with the initial implementation of the project.

References

1. RFC 1493. http://www.it.kth.se/docs/rfc/rfc1493.txt.
2. Cloudscape. http://www.cloudscape.com.
3. DataBahn. http://databahnsoft.com.
4. Oracle Lite. http://www.oracle.com/mobile.olite.
5. Active Networks. http://www.sds.lcs.mit.edu/darpa-activent.
6. Archive of the NMRG Mailing List.
   http://www.ibr.cs.tu-bs.de/projects/nmrg/marchive-1999.
7. The Jaguar Project.
   http://www.cs.cornell.edu/database/jaguar/jaguar.html.
8. The Lore Project, http://WWW-DB.Standford.EDU/lore.

A Transactional Approach to Configuring Telecommunications Services

Tim Kempster, Gordon Brebner, and Peter Thanisch

Institute for Computing Systems Architecture
Division of Informatics,
University of Edinburgh,
Mayfield Road, Edinburgh, EH9 3JZ, Scotland;
Tel +44 131 650 5139, Fax +44 131 667 7209.
This research was supported by EPSRC grant GR/L74798.
{tdk,gordon,pt}@dcs.ed.ac.uk
http://www.dcs.ed.ac.uk

Abstract. The trend in the telecommunications industry towards the provision of IP-based services over large-scale networks creates new challenges in network management. In particular, network managers require the ability to flexibly reconfigure the network. This necessitates enhancements to existing network management techniques, and in particular we require a mechanism for enforcing the atomicity of reconfiguration changes that is robust in the face of network failures. Fortunately, atomic commit protocols have already been developed for achieving atomicity in distributed transaction processing. We demonstrate how this technology can be transferred to network configuration management. We also examine the ways in which atomic commit protocols can benefit from new telecommunications services.

1 Introduction

For many years, industry analysts have been announcing the 'imminent' integration of voice and data network services. After several false dawns, it appears that there may be a real trend in this direction, partly driven by the increasing deployment of IP-based services in the telecommunications industry. Telecommunications providers now route voice, data, fax and broadband services such as video over IP networks. As this trend continues telecommunications networks and IP networks will become indistinguishable, however new challenges in network management will emerge. Network management in this new environment will require the configuration of hundreds of complex routers and switches which may be geographically distributed across thousands of miles. Traditional network management techniques, particularly in the field of network configuration, will require revision to cope with both the larger scale and the complexity of networks, as well as the enriched set of configuration parameters within each network component.

The protocols required to achieve this new level of service will need to embody functionality similar to the *atomic commit protocols* found in distributed

W. Jonker (Ed.): Databases in Telecommunications, LNCS 1819, pp. 40–53, 2000.

database technology[3]. Commit protocols are best known in their role of enforcing the *atomicity* of distributed database transactions: they ensure consistent termination of distributed transactions in the face of site and network failures. Although network configuration can benefit from the use of database commit protocols, leveraging ideas founded in database research, these protocols need to be adapted to the different performance and failure regime of network management applications.

To reinforce the significance of the overlap between these areas, we also review research on new applications for commit protocols in the provision of network services such as atomic broadcast. Finally, we point out ways in which the efficiency of commit protocols can be improved by exploiting new telecommunications services.

2   Setting the Scene

2.1   Network Nodes

Networks are comprised of *nodes* which can be thought of as points of connection together with a *transport medium* connecting these nodes, along which data propagates. Nodes are either the *end points* at which data originates or is consumed, or *redistribution nodes* which store, possibly duplicate or filter, and then forward data to other nodes. Examples of redistribution points are routers, gateways, switches and hubs. These entities can often be configured to change redistribution behaviour and monitored to give a view of data passing through them. The extent of monitoring and configuration that is provided depends on the device. In general however configuration and monitoring capabilities are becoming more comprehensive. One example of this is the advent of the intelligent router[16]. In this paper we will use the generic name *network node* (NN) to describe any redistribution node that can be monitored or configured.

Networks are controlled and monitored from (inter)Network Operation Centers (NOCs)[1]. From a central point an operator gathers information on the state of NNs and the traffic patterns within the network.

Much software has been developed to assist the task of network management. Typically it consists of agents that execute on each NN together with a front end management system which executes at the NOC. HP OpenView's Network Node Manager provides a good example of this. Other vendors have produced interfaces to store network traffic statistics in databases and even analyse this traffic using OLAP style queries. Most software development and research has focused on monitoring rather than control and configuration. In this paper we are interested in the control and reconfiguration of a network service rather than its monitoring.

---

[1] Many NOCs may control a single network or subset of a network. Jurisdiction and arbitration of control is an issue we will not discuss.

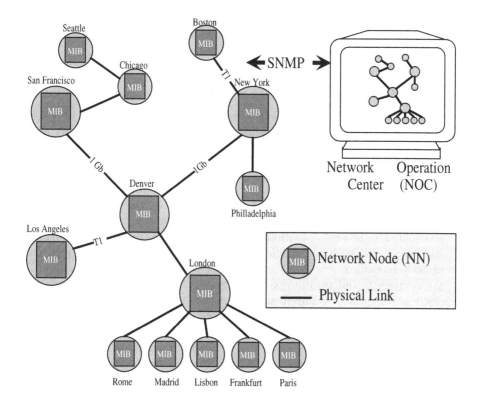

Fig. 1. An NOC communicates with the network using SNMP

2.2   Communicating with Network Nodes

The *de facto* standard protocol used for monitoring state and issuing commands
to change NN behaviour is the simple network management protocol (SNMP)[4].
This protocol utilises the idea of a Management Information Base (MIB)[10][18].
The MIB describes a set of useful configuration, control and statistical objects.
Example objects include routing tables and packet arrival rates. Each object
has an access attribute of either **none**, **read-write** or **write-only**. Each NN
supports some subset of objects in the MIB called SNMP MIB view. Figure 1
sets the scene. In essence SNMP supports three types of operations.

1. **GetRequest** used to retrieve the value of a MIB object from an NN.
2. **SetRequest** used to set the value of a MIB object in an NN.
3. **traps** enables a NN to generate messages when certain events occur (eg. a
   restart).

In this paper we are not concerned with traps and will concentrate on the first
two methods.

   SNMP need not be implemented on top of a guaranteed message delivery
protocol. Monitoring and control software built on SNMP must therefore be

resilient to packet loss. SNMP is usually implemented on top of an unreliable, packet-based protocol such as UDP. Each SNMP request or response requires one packet. We will see this raises some interesting issues later.

2.3   Enriching SNMP Services

SNMP allows very simple operations that update and read objects in a NN's MIB. We argue that by slightly extending these operations[2] we acquire much greater power and flexibility. We are motivated by the following two inadequacies of the current implementation.

P1 SNMP does not allow for provisional changes to be requested. Once a request is made it is either granted or denied. If granted by the time the requester is informed of the change it has been applied. In this case it can only be reversed if an inverse operation exists. For the configuration of very simple services this is perhaps sufficient but as NNs become more complex it is unlikely to suffice.

We envisage reconfiguration problems where the proposed change cannot be easily undone but it is important to test the feasibility of this change before applying it. These types updates are very common in other services . In airline reservation systems, seats can be held before they are bought.

P2 The SNMP `GetRequest` method does not enforce persistent views of objects. In other words once an object in a MIB at a NN has been read that object may be immediately updated. It is often very useful to lock out updates once an object has been read for some period of time. We will see later that this allows us to assert a property known as *serializability* in our future network management schemes.

In order to carry out an update to a MIB object the acquisition of a *resource* may be required. For example, suppose router level filtering of certain types of traffic is to be enabled on a particular port. If this is to be enabled, additional CPU may be required to examine packets in more detail for the filtering to take place. If enough CPU resource is present then the change can be made, if not the request must be denied.

Many systems provide mechanisms where values can be *locked*[3] The notion of locking can be applied to MIB objects to solve the problems **P1** and **P2** above.

The problem **P1** of provisional updates to configuration can be solved in the following way.

UI A provisional configuration is made by sending a `ProvSetRequest` message to the NN. The NN tries to acquire an exclusive lock on the object to be updated. Once the lock has been obtained a provisional update is made and an acknowledgement is sent to the requester.

---

[2] We could build these extensions on top of the existing SNMP methods

[3] The simplest implementation of locking provides two types of locks, exclusive and shared. A shared lock can always be granted unless an exclusive lock exists. An exclusive lock can only be granted if no other lock exists.

UII   Step UI is repeated for all the changes that are required on a NN and then a `PrepareRequest` message is sent to the NN. The NN can now decide if it can carry out the cumulative changes requested. If so the required resources for all the changes are reserved[4], and a `Yes` vote is returned to the requester. If the changes cannot be made a `No` vote is returned. Once a `Yes` vote has been made the NN *must* make the changes requested of it, if later asked to commit. At any time up to the time the `Yes` vote is sent the NN may discard all the updates, release locks and respond with a `No` vote when a `PrepareRequest` arrives.

UIII Once the `Yes` vote is received the requester can decide whether or not to go ahead with the request. If so, it sends the NN a `CommitRequest` to commit the changes requested or else an `AbortRequest`. On receiving an `AbortRequest` the NN may release any locks and resources it tentatively acquired. On receiving a `CommitRequest` it makes the changes and then releases any exclusive locks held.

When requesting the values of MIB objects a similar protocol is used.

R I   A value is requested by sending a `LockingGetRequest` message to the NN. The NN tries to acquire a shared lock on the object. Once a lock has been acquired the value can be returned to the requester.

R II Step R I is repeated until all the read requests have completed. The requester then sends a `PrepareRequest` message to the the NN. This informs the NN that the value no longer need be protected from updates and the shared lock can be released. The NN acknowledges this release with a `Yes` vote. If the lock could not be held the NN can send a `No` vote to the requester to invalidate the value previously returned.

In the example of Figure 2 the NOC first reads a value from an NN in London then provisionally updates a value on a NN in London based on the value read at Frankfurt. The NOC prepares, and then later commits the changes at the London NN. After this a `prepareRequest` is sent to Frankfurt and the read locks are released there.

3   A Centralised Two-Phase Commit Solution

Perhaps the most popular and prevalent commit protocol is the Centralised Two-Phase Commit Protocol[1]. We will describe it in the environment of Network Management. The protocol is comprised of two parts,[5]the operational part and the commit part. During the operational part the operations of the reconfiguration are carried out. This might involve getting object values from MIBs around the network and tentatively updating MIB objects using the methods described

---

[4] Locking at the resource level may also be used to implement this reservation but this should not be confused with MIB object locking.

[5] Strictly speaking, the commit protocol only pertains to the second part.

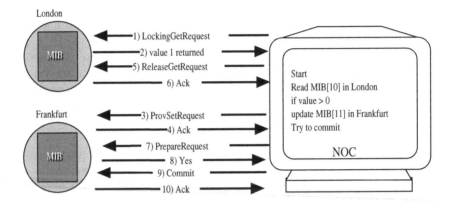

Fig. 2. An NOC performs enriched operations.

in Section 2.3. Once all the operations have been carried out[6] the protocol enters its commit phase.

The centralised commit protocol requires a coordinator for the commit phase. We will assume the same process at the NOC that issued the operations of the transconfiguration also acts as the coordinator. In fact the role of coordinator can be handed over to any process located anywhere on the network. The coordinator solicits votes from each of the NN participating in the transconfiguration using the **PrepareRequest** message. The coordinator collects these votes and then decides on an outcome. If any NN voted **No** or if a timeout expires before all votes are collected the coordinator decides Abort and sends a **AbortRequest** message to any site where a tentative update was performed. If all sites vote **Yes** a **CommitRequest** messages is sent to all non read-only sites. Figure 3 shows the state diagrams of a Coordinator, a NN performing reconfiguration and a NN involved only in read requests.

4   Atomicity and Serializability

The updates and read operations issued at a NOC in order to carry out reconfiguration of the network are analogous to a distributed transaction in the sense that these operations must be grouped together as an atomic unit. The operations of the reconfiguration can be grouped together into a transaction. We call this transactional-style reconfiguration a *transconfiguration*. If implemented carefully we can ensure that the operations within the transconfigurations are atomic.

---

[6] The feasibility of the transconfiguration may already be established in the operational part if a NN notified the NOC early that the change is impossible. In this case the protocol will terminate here with an aborted outcome and will not enter the next commit part.

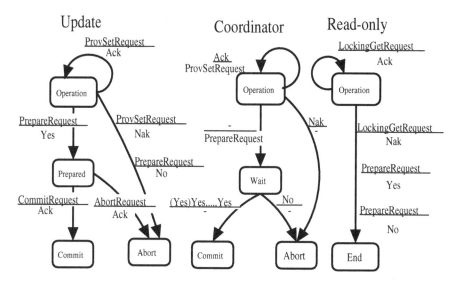

Fig. 3. State diagram of the centralised two-phase commit protocol. Transitions are labeled $\frac{\text{rcv msg}}{\text{snd msg}}$ to represent sending 'snd msg' when 'rcv msg' is received.

The atomicity of a transconfiguration is a very powerful tool for the network manager. The network manager need only specify the operations to reconfigure the network service. The atomicity ensures either all or none of the operations are performed. If one or more NNs cannot fulfill the network manager's requests, the state of the whole network will remain unchanged. Later, we discuss another powerful property of transconfigurations, namely serializability, but for now we will focus on atomicity.

4.1   Atomicity

Thanisch[17] distilled four general properties that are necessary before the use of a commit protocol to enforce transaction atomicity is applicable. These are restated below in Table 1.

Atomic reconfiguration of NNs in a network management environment fulfills all the requirements above. The multiple agents are the NNs. They are semi-autonomous. In general a change may be difficult to undo and finally with the extended operations described in Section 2.3 it is possible to establish the tentative feasibility of their reconfiguration. We can conclude that a commit protocol is appropriate for establishing the atomicity of reconfiguration amongst the NN in a network management environment.

Table 1. The MTSP conditions that presage the use of a commit protocol to ensure atomicity.

M	Multiple Agents.
T	Tentativeness: Agents can tentatively establish feasibility of the change.
S	Semi-Autonomous Agents: they can always be forced to abort, but they can only be forced to commit if they voted to commit.
P	Permanence: an Agent that is required to make changes cannot easily undo these changes.

4.2   Serializability

Serializable level isolation[7] (serializability) arises by combining MIB object locking with a commit protocol. If transconfigurations are serializable a network manager can be sure that if the operations of two transconfigurations execute concurrently on a network, then the effect on the network will be equivalent to running them in some serial order. This is a very useful property especially when multiple network managers have shared control of a network. To illustrate this point further we consider the concurrent executions of two transconfigurations without the property of serializable isolation.

Example 1. Two routers A and B straddle networks N and M routing two types of traffic, x-traffic and y-traffic, from LAN N to LAN M. The network is initially configured so that A routes x-traffic from N to M while blocking y-traffic. Similarly, B routes y-traffic while blocking x-traffic. Figure 4 describes the situation. It is important that each type of traffic is routed from N to M exactly once. Periodically a transconfiguration T (see Figure 4) is executed at a NOC. The transconfiguration toggles routing of the different traffic types between routers.

Suppose two transconfigurations of the type T above, denoted T1 and T2, are run concurrently from different NOCs controlling the same network. If isolation is not maintained the operations of the two transconfigurations could be performed as follows. T1 reads A's routing into variable a (x-traffic). T2 reads A's and B's routing configuration into its variables a and b (x-traffic, y-traffic respectively). T2 then updates A and B with the contents of b and a respectively (setting them to y-traffic and x-traffic respectively). Finally T1 reads the value of B's routing (x-traffic) into its variable b and updates routers A and B with the contents of its variables a and b which both contain x-traffic. The result is that the network is configured to route only x-traffic from N to M. Clearly this is incorrect.

The fundamental serialization theorem[1] states that if a transaction performs locking in a growing phase (where locks are only acquired) followed by a shrinking phase (where locks are only released), then serializable isolation is achieved[8].

---

[7] A transaction can request an Isolation level to limit how much interference is allowed between it and other concurrent transactions[3].

[8] This style of lock acquisition and release in a transaction is often referred to as *well formed two-phase locking.*

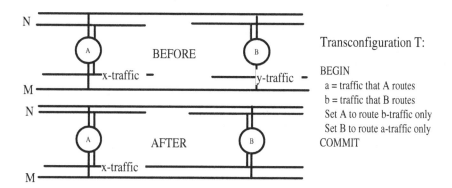

Fig. 4. Transconfigurations interfere producing unexpected results.

In distributed transaction such as our transconfigurations we require barrier synchronization to provide a window between the two phases. Commit protocols provide just that window. In our centralised two-phase commit protocol locks are acquired up to the point the coordinator sends a **PrepareRequest** message. After that point locks are never acquired only released. We can conclude that our transconfigurations will maintain serializable isolation. If serializability is not important then our transconfigurations need never hold read locks, atomicity will still hold.

5 Blocking

The two-phase commit protocol suffers from blocking. Blocking occurs when a NN is prevented from deciding commit or abort due to inopportune failures in other parts of the system[1]. In the centralised two-phase commit protocol if a NN has voted **Yes** and then loses contact with the coordinator it cannot release the locks and resources it has reserved until it establishes contact with the coordinator once more[9]. Since our NOC might rely on the network it is controlling to pass the messages of the commit protocol, failure of a NN might cause the network to partition leaving the coordinator in one partition and the NN in another. If this was to happen after a NN had voted **Yes** but before it was informed of the outcome the NN would block. Protocols have been designed that are more resilient to failure and we will discuss the appropriateness of these in next.

The three-phase commit protocol[13] is often proposed to solve the problem of blocking. However this proposal needs qualification. In environments where networks can partition and messages can be lost three-phase commit also suffers

---

[9] Strictly speaking, if it were able to contact another NN involved in the transconfiguration that had either been informed of the outcome or had not yet voted it could resolve the transconfiguration.

from blocking. In fact, there exists no protocol resilient to a network partitioning when messages are lost[15]. Although no non-blocking protocols exist the three-phase commit protocol has undergone extensions which guarantee that after a network partition, if any subset of NNs form a connected quorum[10], this subset will not block. The first such protocol was Skeen's quorum based three-phase commit[14]. This was further extend so that nodes could survive repeated partitioning[5] and then again extended to produce commit outcomes more of the time[6].

The choice on whether or not to adopt a quorum based three-phase commit protocol depends on two factors. The first is connectivity. Often an NOC will have the ability to establish a dialup connection with a NN should it become isolated from the NN by network failure in the network it is configuring. If this is the case then a centralised approach with the NOC acting as coordinator will be appropriate. The second factor is the ability of the NNs to elect a coordinator in a partitioned network. For a quorum based protocol to be useful the partitioned NNs must be capable of electing a coordinator and carrying out a termination protocol to decide an outcome. Currently NNs cannot be configured to perform this but in the future this might be possible.

6 Special Requirements for Network Management

In this section we discuss some special requirements that network managers might have and how current database research can address these requirements. We also look at a common two-phase commit optimization *early prepare* which is particularly appropriate for network management style commit processing.

6.1 Real-time Commit Protocols

Although the commit protocols we have discussed so far guarantee atomicity they do not guarantee that the changes will take place simultaneously. If this were possible then the well known impossibility of clock synchronization[9] in a distributed environment with unpredictable message delay would be contradicted. The best we can hope for therefore is that the changes take place within some time window.

Once again the database community has addressed this problem when considering commit processing in Real-Time Distributed Database Systems (RT DDBS). These systems handle workloads where transactions have completion deadlines. Davidson et al.[2] describe a centralised two-phase commit protocol where the fate of a transaction is guaranteed to be known to all the participants of the transaction by a deadline, when there are no NN, communication or clock faults. In case of faults, however, it is not possible to provide such guarantees and an exception state is allowed which indicates the violation of a deadline.

---

[10] The simplest example of a quorum is a majority.

Using a real-time commit protocol a network manager can specify a deadline at the point he initiates the commit protocol. In this way he can be sure that the changes will take effect within a specific time window, minimising disruption to the network.

6.2   Order of Commitment

It might be important for a network manager to specify an order[11] in which changes made by a transconfiguration should be applied. This might be important to stop cycles in traffic flow in the network which could result in network storms. We discuss two techniques from the world of transaction processing which allow us to impose such an order.

If a centralised two-phase commit protocol is being used where changes are not applied until commit time the order of commitment can be controlled by the coordinator. The coordinator can dispatch each `CommitRequest` in order waiting for acknowledgements before proceeding to the next request.

Nested transactions[3] provide a mechanism where a transaction can be represented as a tree structure. For an parent transaction to commit each child must be willing to commit. In these schemes a strict the order of commitment is often imposed based on the structure of the nested transaction.

6.3   Early Prepare and Logging

If commit protocols are to be resilient to failure certain state changes must be recorded on stable storage by the participants of a protocol. If a participant crashes, upon recovery a *recovery protocol* executes and utilises this stored information to complete the protocol.

Much work has been carried out to reduce both the number of log writes and the number of messages which must be sent during a commit protocol[12]. These optimizations (particularly those which reduce log writes) are perhaps not so important in a Network Management scenario but we will briefly discuss the Early Prepare[11] optimization[12] which reduces the message overhead of the centralised two-phase commit protocol.

In this optimization once a NN has tentatively performed its changes it enters its prepared state and votes `Yes`. This saves an extra message round as the coordinator need not solicit the NNs vote and wait for a reply. This optimization is particularly appropriate where all the update requests can be made to a NN in one batch. If however the NOC requires a dialogue with a NN the method can be expensive because forced write to stable storage is required when (re)entering the prepared state.

---

[11] In general a partial order may be more appropriate but for simplicity we assume a total order.

[12] Early prepare is sometimes referred to as Unsolicited Vote

7   Related Work

Several other proposals have been made for deploying commit protocols as part
of a telecommunications service and we give three examples here.

7.1   Atomic Broadcast

Luan and Gligor's atomic broadcast algorithm [8] has each site (end point nodes)
accumulating received broadcast messages in a queue. When an application at
one of the sites wants to consume a prefix of these messages, it initiates a three-
phase commit protocol in which the other sites vote according to whether they
have received all messages up to and including that prefix. Message sequences
are only consumed by the application if a quorum of sites confirms that messages
up to and including that sequence have been received.

The advantage of this algorithm is that it controls the flow of protocol messages that request progress in the consumption of broadcast message streams. The
protocol is designed to detect the situation where two or more sites have concurrently initiated the protocol. It can then combine the voting of the hitherto
separate protocol executions.

7.2   Network Control for Mixed Traffic

Li *et al.* have proposed the deployment of a commit style protocol in network
management in order to facilitate connection setup in switching systems supporting mixed circuit and packet connections for broadband service [7]. This is part
of a control architecture that provides efficient connection setup in a switching
system that supports mixed circuit and packet connections for broadband services. Their control architecture is intended for use in systems which allow only
one connection to an output port at any given time.

A variety of connection styles, e.g. broadcast, multicast and conferencing,
might be demanded by various users. Li *et al.* note that for configuring some of
these services, a centralised system is preferable, whereas for other services, a
distributed arrangement may be more efficient. Their proposed scheme, using a
commit-like protocol, is an attempt to get the best of both worlds.

The role of the protocol is to keep the port allocation status information
consistent in the various controllers.

Li *et al.* propose a multilevel hierarchical architecture in which Virtual-
Centralized Controllers (VCCs) are attached to each level of the control network.
The current status of each port can be obtained from each VCC.

Phase 1: (Connection Request Processing). Each VCC, VCC_i, makes a tentative
decision on whether the connection requests can be granted, based on the port
status at the end of the previous cycle. A tentative allocation, T_i, is made by
VCC_i.

Phase 2: (Tentative Allocation Broadcast Phase). Tentative decisions are sent
up and down the hierarchy. Thus each VCC collects the tentative decisions made
by its parents and children.

Phase 3: (Conflict resolution Phase). Conflicts that occurred in the tentative allocations are resolved and a final allocation is generated. Various conflict-resolution rules are possible, typically assigning different priorities to different levels.

None of the ports on a destination list of a multicast or a broadcast connection will be allocated unless *all* of them are available.

Li *et al*'s protocol works because they are able to assume that the protocol is completely synchronised: the second phase ensures global delivery.

7.3   Active Networks

Zhang *et al.* [19] have investigated the possibility of exploiting an active network in commit protocol execution. They propose to have the active network nodes collecting and examining protocol votes. One active node is designated as the coordinator. Other active nodes can suppress duplicated votes sent towards the coordinator node in a hierarchical network.

We note that such network facilities could be exploited by many of the commit protocols we have discussed in this paper.

8   Conclusions

We have identified a potentially useful area of overlap between research in telecommunications services and research in distributed transaction processing. Telecommunications service providers can adapt transaction processing protocols to the new application of network configuration management and those developing more efficient commit protocols can exploit new telecommunications services. However, the performance and robustness properties of these technologies could potentially limit the extent of this symbiosis.

References

1. P.A. Bernstein, V. Hadzilacos, and N. Goodman. *Concurrency Control and Recovery in Database Systems*. Addison-Wesley, Reading, MA, 1987.
2. S. Davidson, I. Lee, and V. Wolfe. A protocol for timed atomic commitment. In *Proceedings of the Ninth International Conference on Distributed Computing Systems*, pages 199–206, Newport Beach, CA, June 1989. IEEE Computer Society Press.
3. J. Gray and A. Reuter. *Transaction Processing: Concepts and Techniques*. Morgan Kaufmann, San Mateo, CA, 1993.
4. M. Schoffstall J. Case, M. Fedor and J. Davin. A simple network management protocol (snmp). Technical Report Internet Draft IETF 1157, IETF, May 1990. On-line, http://www.ietf.cnri.reston.va.us/rfc/rfc1157.txt.
5. I. Keidar and D. Dolev. Increasing the resilience of atomic commit, at no additional cost. In *Proceedings of the 14th ACM SIGACT-SIGMOD-SIGART Symposium on Principles of Database Systems*, pages 245–254, 1995.

6. T. Kempster, C. Stirling, and P. Thanisch. More committed quorum-based three phase commit protocol. In *Lecture Notes in Computer Science: The Twelth International Symposium on Distributed Computing*, page 246, 1998. On-line, http://www.dcs.ed.ac.uk/home/tdk/.

7. C.-S. Li, C.J. Georgiou, and K.W. Lee. A hybrid multilevel control scheme for supporting mixed traffic in broadband networks. *IEEE Journal on Selected Areas in Communications*, 14(2):306–316, February 1996.

8. S. Luan and V.D. Gligor. A fault-tolerant protocol for atomic broadcast. *IEEE Transactions on Parallel and Distributed Systems*, 1(3):271–285, July 1990.

9. J. Lundelius. Synchronizing clocks in a distributed system. Technical Report MIT-LCS//MIT/LCS/TR-335, MIT, Massachusetts Institute of Technology, Laboratory for Computer Science, August 1994.

10. K. McCloghrie and K. McCloghrie. Management information base for network management of tcp/ip-based internets. Technical Report Internet Draft IETF 2246, IETF, May 1990. On-line, http://www.ietf.cnri.reston.va.us/rfc/rfc2246.txt.

11. G. Samaras, K. Britton, A. Citron, and C. Mohan. Two-phase commit optimisations and trade-ffs in the commercial environment. In *Proceedings of the 9th International Conference on Data Engineering*, pages 520–529, April 1993.

12. G. Samaras, K. Britton, A. Citron, and C. Mohan. Two-phase commit optimisations in a commercial distributed environment. *Distributed and Parallel Databases*, 3(4):325–360, October 1995.

13. D. Skeen. Nonblocking commit protocols. In *Proceedings of the ACM SIGMOD Conference on the Management of Data (SIGMOD'81)*, pages 133–142, 1981.

14. D. Skeen. A quorum-based commit protocol. In *Berkeley Workshop on Distributed Data Management and Computer Networks*, pages 69–80, February 1982.

15. D. Skeen and M. Stonebraker. A formal model of crash recovery in a distributed system. *IEEE Transactions on Software Engineering*, SE-9(3):220–228, May 1983.

16. David L. Tannenhouse and David J. Weatherall. Towards an active network architecture. *Computer Communications Review*, 26(2), April 1996.

17. P. Thanisch. Atomic commit in concurrent computing. Accepted for publication in IEEE Concurrency, 1999.

18. S. Waldbusser. Remote network monitoring management information base. Technical Report Internet Draft IETF 1757, IETF, Febuary 1995. On-line, http://www.ietf.cnri.reston.va.us/rfc/rfc1757.txt.

19. Z. Zhang, W. Perrizo, and V. Shi. Atomic commitment in database systems over active networks. In *ICDE'99*, pages –. IEEE Comput. Soc. Press, 1999.

Making LDAP Active with the LTAP Gateway

Case Study in Providing Telecom Integration and Enhanced Services

Robert Arlein, Juliana Freire, Narain Gehani,
Daniel Lieuwen, and Joann Ordille

Bell Labs/Lucent, 700 Mountain Ave, Murray Hill, NJ 07974, USA,
{rma,juliana,nhg,lieuwen,joann}@research.bell-labs.com,
WWW home page: http://www-db.research.bell-labs.com/index.html

Abstract. LDAP (Lightweight Directory Access Protocol) directories are being rapidly deployed on the Web. They are currently used to store data like white pages information, user profiles, and network device descriptions. These directories offer a number of advantages over current database technology in that they provide better support for heterogeneity and scalability. However, they lack some basic database functionality (*e.g.,* triggers, transactions) that is crucial for Directory Enabled Networking (DEN) tasks like provisioning network services, allocating resources, reporting, managing end-to-end security, and offering mobile users customized features that follow them. In order to address these limitations while keeping the simplicity and performance features of LDAP directories, unbundled and portable solutions are needed.

In this paper we discuss LDAP limitations we faced while building an LDAP meta-directory that integrates data from legacy telecom systems, and how LTAP (Lightweight Trigger Access Process), a portable gateway that adds active functionality to LDAP directories, overcomes these limitations.

1 Introduction and Motivation

LDAP (Lightweight Directory Access Protocol) directories [16,33] are being rapidly deployed on the Web. They are currently used to store data like white pages information, user profiles, and network device descriptions. Compared to current databases, LDAP directories have better support for handling scale and heterogeneity [18]. However, their lack of support for standard database functionality such as triggers and concurrency control hampers their use in more sophisticated tasks, such as Directory Enabled Networking (DEN) services.

DEN is an enabling technology for a wide variety of tasks, including reporting, managing end-to-end security, and offering mobile users customized features that follow them [15]. While this technology is not limited to LDAP directories or to any particular standard, it is frequently associated with the effort by equipment and software vendors to standardize LDAP schemas to support Directory

W. Jonker (Ed.): Databases in Telecommunications, LNCS 1819, pp. 54–73, 2000.
© Springer-Verlag Berlin Heidelberg 2000

Enabled Networking. The DEN standardization proposes the use of a single logical source for needed information and encourages storage of data in a standard format to enable interoperability and value-added reseller innovation.

As part of Lucent's DEN initiative, we built the MetaComm system. MetaComm integrates data from multiple devices into a meta-directory, allowing user information to be modified through the directory using the LDAP protocol as well as directly through two legacy devices: a Definity® PBX and a voice messaging system. In order to prevent data inconsistencies, updates to any of the systems must be reflected appropriately in all the other systems. For example, when a new employee is hired and her information is added to the directory, changes to a variety of devices are made automatically. By providing a simpler, unified interface to data stored in telecom devices, MetaComm greatly simplifies access to this data, reduces the need to manually re-enter data in multiple devices, and also reduces data inconsistencies across repositories.

In order to provide such functionality, MetaComm needs to detect relevant changes to data stored in the directory server in a timely manner. A natural solution would be to have triggers in the server that notify MetaComm of changes. However, there is no standard for LDAP triggers[1], and trigger functionality is not currently available in many commercial servers. In the absence of triggers, one alternative would be to have MetaComm poll the directory to *discover* changes; in our application, however, this is not practical given the resources required and delays it generates.

Our challenge was then to come up with a solution that is portable (*i.e.,* works with any LDAP server) and efficient. Existing approaches to add trigger functionality to database and directory servers require either:

1. modification to system internals (*e.g.,* adding either a full-fledged trigger system or associating plug-ins with directory commands),
2. knowledge of proprietary log formats, or
3. periodic polling of database state.[2]

In MetaComm we used LTAP (Lightweight Trigger Access Process)[3], a gateway that adds active functionality to *any* LDAP server. LTAP does not require the use of any of the above methods – it adds active functionality to LDAP without requiring any proprietary extensions to the protocol, a key advantage. Thus, LTAP can be used to improve interoperability and portability. Another advantage of LTAP is that by unbundling triggers, users that do not need trigger functionality need not incur any added overheads. For instance, a read-only replica of MetaComm's LDAP directory does not need to be front-ended by LTAP.

---

[1] The IETF LDAP Extension Group, ldapext, has on-going work to standardize LDAP triggers [22].

[2] The proposals on persistent search at the ldapext group will supply continuous polling of database state efficiently. Once available, persistent search will provide an attractive platform for implementing triggers that fire after an update has been executed. However, persistent search has limitations, for example, it cannot be used to build triggers that fire before an update is executed.

[3] LTAP can be downloaded from `http://ltap.bell-labs.com`.

```
objectclass person
        oid 2.5.6.6
        superior top
        requires
                sn,
                cn
        allows
                description,
                seeAlso,
                telephoneNumber,
                userPassword
```

Fig. 1. Definition of the X.500 `person objectclass`

In what follows we describe our experiences with using LDAP in a meta-directory, its limitations, and our approach to overcome them. The paper is organized as follows. Sect. 2 contains an overview of the LDAP protocol. Sect. 3 describes the overall architecture of MetaComm. Sect. 4 describes the LTAP gateway and its functionality by way of an extended example based on triggers used in MetaComm. Sect. 5 summarizes our experiences in building MetaComm, and the implications of our gateway approach for adding needed but missing functionality to LDAP. These experiences are key lessons learned in this paper. They will be valuable to those building similar systems. Sect. 6 contains related work, and conclusions and future work are in Sect. 7.

2   LDAP (Lightweight Directory Access Protocol) Overview

LDAP is a widely deployed directory access protocol with implementations by a large number of vendors (see reference [17] for a partial list). LDAP can be thought of as a very simple database query and update protocol. From a database perspective, LDAP has some beneficial properties, for example, it deals well with heterogeneity and allows highly distributed data management while keeping data conceptually unified [7].

LDAP servers typically fix a schema at start up time. Each *directory entry* (analogous to a tuple in a relational database) has at least one type – called an `objectclass` – associated with it. An entry can have multiple types, and types can be added to or deleted from the entry at run time. An example of an `objectclass` is the X.500 `person` class in Fig. 1. Sub-typing is provided, for instance, `person` inherits from the `top` class. Note that certain fields must be specified for each `person` (*e.g.*, `sn` – surname, `cn` – common name), while others are optional (*e.g.*, `description`, `telephoneNumber`). All attributes are strings that are "weakly typed." For instance, a `telephoneNumber` is a special kind of string, a tel[4]. One can assign any value to a `telephoneNumber`, because no type checking is performed for assignments. Nonetheless, when the system compares two `tel` values, it compares them in the expected way. For instance, ''908-582-5809'' and ''9085825809'' are considered equal. Similarly, common name (`cn`) is a case insensitive string (`cis`), thus, when it is compared with another string, case is ignored. In addition, attributes are multi-valued, for example, a `person` may have multiple `telephoneNumbers`.

Directory entries are stored in a tree or forest. Fig. 2 is an example of a typical tree, simplified to remove all but one attribute from each entry. Each entry in the tree is identified by a *Distinguished Name* (DN) which is a path from the root of the tree to the entry itself. The DN is produced by concatenating the *Relative Distinguished Name* (RDN) of each entry on the path. The RDN for an entry is set at creation time and consists of an attribute name/value pair – or in more complicated cases, a collection of these pairs connected with various operators. The RDN of an entry must be unique among the children of a particular parent entry. When using Fig. 2 in examples, we assume that the attribute/value pairs in the figure (*e.g.*, ''o=Lucent'' and ''cn=John Doe'') are the RDNs. Thus, for example, the DN for John Doe is ''cn=John Doe, o=Marketing, o=Lucent''. Note the leaf-to-root order is the reverse of that for the representation of a UNIX® file or a URL.

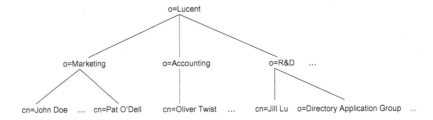

Fig. 2. Sample LDAP Tree

The only update commands in LDAP are to create or delete a single leaf node or to modify a single node. There are two kinds of modification commands. The first can modify any fields except those appearing in the RDN. The second, ModifyRDN, cannot change any attribute/value pairs except those appearing in the RDN. Furthermore, while individual update commands are atomic, one cannot group several update commands into a larger atomic unit (*i.e.*, a transaction). For instance, one cannot atomically change a person's name and telephone number if the name is part of the person's RDN but the telephone number is not. In addition to providing limited update capabilities, LDAP provides query facilities based on the structure of the tree.

---

[4] The typing of the fields is specified separately in the schema, and all attributes with the same name have the same type.

3   MetaComm: A Telecom Meta-Directory

A great deal of corporate data is buried in network devices such as PBXs, messaging/email platforms, and data networking equipment, where it is hard to access and modify. Typically, the data is only available to the device itself for its internal purposes and it must be administered using either a proprietary interface or a standard protocol against a proprietary schema. In addition, since some of this data is needed in multiple devices as well as in applications like corporate directories and provisioning systems, it is often re-entered manually and partially replicated in different devices in quite different formats. As a result, data is not only hard to access, but there is also a great deal of expense involved in performing updates and the risk of inconsistencies across devices.

The MetaComm system is part of Lucent's DEN initiative. MetaComm integrates data from multiple devices into a meta-directory, allowing user information to be modified through a directory using the LDAP protocol as well as directly through two legacy devices. In order to prevent data inconsistencies, updates to any of the systems must be reflected appropriately in all the other systems. In this scenario, maintaining consistency poses many challenges, because data integration is performed across several systems with no triggers and with extremely weak typing and transactional support. LTAP is used as infrastructure to solve these problems, providing features like locking and detecting LDAP updates.

In what follows we give a brief overview of the system, how it uses LTAP triggers, and the applications it enables. For a detailed description of MetaComm the reader is referred to [10].

3.1   MetaComm: Architecture and Features

The first prototype of MetaComm integrates user data from two legacy devices: a Definity® PBX and a messaging platform. As shown in Fig. 3, our implementation allows user data to be modified in two ways. First, the data can be modified through an LDAP directory which materializes the data from legacy devices. Second, users can continue to modify the telecommunication devices directly through existing proprietary interfaces. Offering multiple paths to modify parameters in crucial telecommunication devices preserves the experience base of administrators of the devices, and it increases the overall reliability and availability of the system since updates can still be made directly to the device even if the directory becomes inaccessible.

Fig. 3 shows the various components of MetaComm. The Update Manager (UM) is the central component of the system – it ensures that the data in the devices and in the LDAP server are consistent. Note that consistency is not just a matter of applying the same update to each data repository in a global transaction. Because the repositories lack most basic transaction facilities, MetaComm uses other techniques to ensure that the repositories converge to the same values after some delay (*e.g.,* [30], [9]). These techniques include re-applying updates on certain repositories to ensure the serializability of updates,

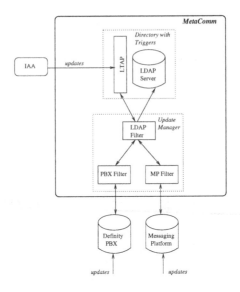

Fig. 3. MetaComm Architecture

and recovery from catastrophic communication or storage errors through re-synchronization of the repositories.

Maintaining the consistency of the repositories also requires that the semantics of the data is properly reflected in each repository. A filter or wrapper is associated with each repository. In MetaComm there are three such filters, the PBX filter, the Messaging Platform (MP) filter and the LDAP filter, depicted in Fig. 3. Each filter has a protocol converter for communicating with its associated repository and a mapper for translating update commands to/from the schema of the repository. The schema translation and integration of the mapper are realized through *lexpress* [28]. lexpress uses semantic characteristics of the data to simplify data integration. In particular, lexpress uses data dependencies to propagate data wherever it is needed in the global or device schema, and partitioning constraints to translate schema updates correctly and route them to the proper repositories.

Each repository in the system, the legacy devices and the LDAP directory server, must notify the UM when a change occurs. The LTAP module adds active functionality to the LDAP server and notifies the UM of changes to data in the LDAP directory. The main thread of the UM responds to update and synchronization requests by propagating update commands to the appropriate filters. The mapper component in the filter further analyzes the request to ensure that updates are properly forwarded to the associated data repository.

Also shown in Fig. 3 is the Integrated Administration Application (IAA), which provides a single point of administration for the telecom devices. It is worth pointing out that any LDAP tool (*e.g.,* an LDAP enabled Web browser) can contact LTAP to administer the telecom devices.

3.2   LTAP: Triggers and Locking

LTAP allows triggers to be defined and associated with directory entries. LTAP accepts LDAP messages and returns standard LDAP replies, so it looks exactly like an LDAP server to the outside world. However, it does not perform storage, updates, or queries. LTAP parses each request to determine whether the request is relevant to its triggers. If a request is not relevant it is simply forwarded to the LDAP server, otherwise LTAP performs the required trigger processing actions.

MetaComm uses LTAP as a gateway for its LDAP server. The integrated schema of MetaComm extends the standard X.500 class [5] that describes people with auxiliary classes that represent device specific information. The materialized view of the integrated information is stored in an LDAP server. LTAP triggers are attached to portions of the schema that represent data shared among the devices, ensuring that MetaComm is notified whenever *relevant* data changes.

Besides triggers, MetaComm uses the locking capabilities provided by LTAP (see Sect. 4.3). As depicted in Fig. 3, an update request originating at a device is translated into an LDAP command (to update the global schema of the LDAP server by the appropriate filter) and forwarded to LTAP by the LDAP filter. Because the PBX, MP and the LDAP server lack locking capabilities, the translated updates are sent to LTAP first, so that the corresponding entry is locked before the update is forwarded to the UM. Locking and the possible re-application of the update to the devices ensure proper serialization order.[5]

A more detailed description of LTAP, its functionality and how it is used in MetaComm, is given in Sect. 4.

3.3   New Applications Enabled by MetaComm

MetaComm allows modification of PBX/messaging settings through any LDAP tool (there are a variety of GUI interfaces to LDAP directories). In our project, we were able to use a commercial LDAP product to easily generate an intuitive Web interface.

Using MetaComm administration, an authorized user/program can easily redirect a telephone extension to a port in another room. Traditionally, this could only be done by a highly trained PBX administrator. For example, in applications like hoteling[6], PBX information is not updated when a different person takes possession of a workspace and its telecom resources. Thus, the hoteling employee's calls interact poorly with caller id, leading to callers being identified with uninformative descriptions like "Hotel 1". Once the mapping from extension to port can easily be changed through LDAP, the same tools that assign workspaces can also fix the mapping. In addition, LDAP browsing tools can be used to find out whether a person has any messages, something that

---

[5] It is worth pointing out that the devices and the directory may temporarily diverge, but they are guaranteed to converge to a common value based on the order in which they are applied by the UM [10].

[6] In hoteling, employees who frequently are outside the office share a pool of workspaces. They reserve a workspace when they need to come to the office.

previously required either a proprietary tool on a PC, making a phone call, or looking at the message light on a phone.

3.4   MetaComm Status

MetaComm was included in a demo at InterOp [27]. The portability advantages of LTAP were demonstrated when preparing the demo since we were able to change directory brands easily. Lucent has announced a product that will use the MetaComm technology (called Directory Synchronization Technology in the press release) to control Definity® PBXs through an LDAP directory. The technology is currently being transitioned and hardened for commercial use.

4   The LTAP Gateway

LTAP uses the standard *Event-Condition-Action* (*ECA*) model [8]. Triggers are specified to fire when a directory entry is *affected*, that is, read or written by an operation. Users specify an *Event* to monitor, a *Condition* to check, and an *Action* to perform if the event occurs and the condition holds. Users specify these by instantiating either a C++ or Java `Trigger` object and then registering it with the system. Each specified event monitors the progress of an LDAP operation, for example an event can be set to trigger before an LDAP modify or after an LDAP delete. The condition must be satisfied by the entry affected by the operation. For example, the condition might specify that the affected entry must be of a particular type and must involve the modification of certain attributes.

An action is implemented by sending a message to a *trigger action server* (TAS) with information about the trigger and the operation that fired the trigger so that the TAS can perform the action. Users must provide the TASs for their applications by adding three functions to an LTAP-provided program shell (see Sect. 4.4 for more details).

Rather than go into detail on the syntax, we give an example to illustrate the basic trigger features. Full details can be found in the LTAP user manual [1]. The example is based on the triggers used by MetaComm (described in Sect. 3). LTAP notifies MetaComm whenever a person entry is added, modified, or deleted. We illustrate the creation and execution of a trigger for modification alerts – similar features are required for adds and deletes.

4.1   Trigger Creation

Trigger creation involves defining both the trigger itself and the trigger action, as well as ensuring that a trigger action server (TAS) that can perform the action is available. A trigger action is specified as follows:

   `Action actionPersonMod(`*ActionMachine, ActionPort*`)`;

where *ActionMachine* is a string specifying a machine to contact and *ActionPort* is an integer specifying a port number. These two parameters specify a TAS. Having specified the action, MetaComm next constructs the `Trigger` in Fig. 4. The parameters have the following meanings:

```
Trigger triggerPersonMod(
    "triggerPersonMod",  /* (1) name of trigger */
    OpModify,            /* (2) Modify causes trigger
                                to fire */
    Before,              /* (3) Trigger fires and returns
                                before Modify executed*/
    actionPersonMod,     /* (4) Execute action specified
                                by actionPersonMod*/
    ScopeType,           /* (5) Trigger based on type of
                                the entry*/
    "person");           /* (6) Trigger placed on person
                                entries */
```

Fig. 4. **Trigger** Definition used in MetaComm

1. The first parameter, *e.g.,* "**triggerPersonMod**", specifies the name of the trigger. It is the key for the trigger and allows LTAP to locate the trigger for modification/deletion.
2. The second parameter, *e.g.,* **OpModify**, specifies the LDAP operation, in this case a modify, that causes the trigger to fire. (The second and third parameters are combined to define the event.)
3. The third parameter, *e.g.,* **Before**, specifies when the trigger firing takes place in relationship to the LDAP operation. In this example, the action is executed **Before** the modify is attempted. This parameter could also specify that the trigger should fire **After** the LDAP operation has successfully completed or after the operation has **Failed**.
4. The fourth parameter, *e.g.,* **actionPersonMod**, specifies the action to execute after the event has occurred if the condition holds.
5. The fifth parameter, *e.g.,* **ScopeType**, specifies an LDAP search scope. The search scope can be a portion of the naming tree, specified by **ScopeBase**, **ScopeOneLevel**, or **ScopeSubtree**, or objects of a particular type, specified by **ScopeType**. The trigger event is monitored at all entries that are returned by an LDAP search with scope specified by the fifth parameter and with the DN or **objectclass** specified by the sixth.
6. The sixth parameter, *e.g.,* "**person**", is either a DN – if the fifth parameter is **ScopeBase**, **ScopeOneLevel**, or **ScopeSubtree** – or an **objectclass** – if the fifth parameter is **ScopeType**. Given our combination of fifth and sixth parameter, this trigger fires before any modifications of **person** entries.

The example in Fig. 4 illustrates only the mandatory parts of the trigger, the event and the action. An LTAP condition is optional and contains two parts: a list of fields to monitor, **triggerMonitoredFields**, and an LDAP filter. If the trigger specifies **triggerMonitoredFields**, the trigger action will not be executed unless at least one field in the monitored list is *involved* in the LDAP command that caused the firing. If **triggerMonitoredFields** is not explicitly set, this is equivalent to specifying a list containing all possible fields that could be modified. For example, in MetaComm triggers are defined to monitor all fields containing data that is used by the various devices (see Sect. 3 for more details).

In addition to preventing triggers from firing when uninteresting fields are modified, `triggerMonitoredFields` serves a security function – LTAP only fires a trigger if the trigger writer is authorized to read at least one of the monitored fields that are involved in the operation (see [24] for details).

The second part of the LTAP condition, the LDAP filter, constrains the trigger to fire only when the object satisfies the filter, for example when the object has a particular value for an attribute. The trigger fires whenever a search requested by the trigger writer and rooted at the specified DN, scope, and filter would return the entry. To continue the example, suppose MetaComm only cares about modifications to a person's entry if the `telephoneNumber` field is set. Then, MetaComm would add the following line:

> `triggerPersonMod.triggerFilter = ''(telephoneNumber=*)'';`

This could be used as an optimization if system rules specify that a person created with no telecom access will not get PBX/messaging services added later.

TASs may specify `triggerFieldsOfInterest`, a list of fields that LTAP needs to include in a trigger notification message. LTAP provides TASs the ability to access the values of fields pre-update (`NeedsBefore`), post-update (`NeedsAfter`) or both (`NeedsBoth`), or specify neither (`NeedsNeither`) when no values are needed.

The TAS for MetaComm does not specify a value for `triggerFieldsOfInterest` since it is interested in all of them, and all fields is the default. However, MetaComm needs information about how the fields are modified and what their pre-update values are in order to properly update the controlled devices. As it needs both the pre- and post-update values, it adds the following line:

> `triggerPersonMod.triggerNeeds = NeedsBoth;`

Having constructed the `Trigger`, MetaComm installs the trigger into the LTAP gateway by including the following line:

> `ldapConnection.createTrigger(triggerPersonMod);`

where `ldapConnection` is a handle created when MetaComm's Update Manager connects to LTAP.

4.2   Trigger Processing

The following steps (illustrated Fig. 5) are performed in basic trigger processing. Note that we do not consider time outs in this discussion, and failure handling is described in the user manual [1].

1 LTAP receives an LDAP command C from a user U for entry E.

2/3 LTAP reads trigger information from the trigger database for E. Note that one instantiation of LTAP caches the trigger database in its entirety in main memory within the LTAP process.

4/5 LTAP reads pre-update information for E (if required by some trigger).

6/7 LTAP processes each matching `Before` trigger, passing relevant information to the corresponding TAS – in practice, there may be several TASs contacted, but for simplicity only one appears in Fig. 5. Note that the `Before` trigger

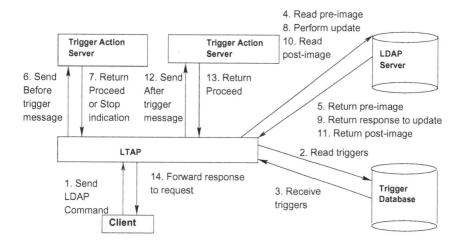

Fig. 5. Interactions during Trigger Processing

creator is required to have super-user status, because the `Before` TAS can assume the access rights of the user. Because access issues are resolved when the trigger is created, no additional run-time authorization is needed. For more details on security issues in LTAP, the reader is referred to [24].

TASs for `Before` triggers return one of three values to LTAP after completing:

1. `Proceed`: instructs LTAP to execute the command C received in step 1.
2. `StopFailure`: instructs LTAP not to execute C and to return failure to U.
3. `StopSuccess`: instructs LTAP not to execute C but to return success to U. An example use for `StopSuccess` is to allow updating through views (*e.g.,* [20]). The original command is not executable against the underlying repository, so sending it there would fail. However, the TAS can rewrite the update and issue the rewritten update itself. If the rewritten update succeeds, the TAS returns `StopSuccess`. The use of `StopSuccess` in MetaComm is discussed in Sect. 5.

14 If any TAS returns `StopFailure` or `StopSuccess` in step 7, then return failure to U if a least one `StopFailure` was received, else return success.

8/9 Execute the LDAP command received in step 1.

10/11 For each After trigger, assume the access rights of the trigger creator and read the update information for E.

12/13 Execute `After` triggers.

14 Return status of LDAP command received in step 9.

It should be noted that LDAP command failure will cause the following change – rather than proceeding from step 9 to step 10, LTAP executes any `Failed` triggers that are authorized to execute and returns the result that was received during step 9 to U.

4.3   Locking

Write locking is required to protect the integrity of the pre- and post-images. Updates to a directory entry while trigger processing on that entry is in progress must be prevented. Thus, when an LDAP update command arrives at LTAP (step 1 in Fig. 5), a write lock is acquired. The lock is held until all the After triggers have been processed (step 13 in Fig. 5). Read requests ignore locks. A write to a single entry is atomic, so each read will see internally consistent entries, *i.e.*, LTAP gives the standard LDAP guarantees for reads. MetaComm, on the other hand, depends on LTAP's locking facilities to maintain consistency between the directory and telecom devices.

Certain users cannot afford to have any operations blocked due to locking. To support such users, LTAP allows locking to be turned off in a configuration file.

4.4   Trigger Action Server Support

In the version of LTAP used in MetaComm, a trigger action contacts a specified port of a TAS. The TAS then executes some code in response to this contact. A TAS returns Proceed, StopSuccess, or StopFailure in response to a Before trigger request.[7] For After/Failed triggers, a TAS simply returns a Proceed as a *done* indication. The LTAP release provides a simple TAS program that can easily be tailored by defining three functions of the TriggerFields class as described below. The user then compiles the TAS and starts it on the machine specified in the trigger.

When the TAS is contacted, a number of items are included in the message. For instance, authorized fields that appeared in the triggerFieldsOfInterest of the Trigger associated with this invocation of the TAS are sent if requested. Each field sent is packaged in an AttributeValue. For each field, an AttributeValue stores its name, attributeName; what happened to it, changeType; and its pre-update (post-update) value, beforeValue (afterValue). Note that the pre-update (post-update) value will only be sent if NeedsBefore (NeedsAfter) or NeedsBoth was specified by the trigger creator. Also, beforeValue and afterValue are arrays of values because fields may be multi-valued in LDAP.

In addition to the pre- and post-update values in the affected entry, the following fields are sent in the message to the TAS:

1. actionDN is the Distinguished Name of the affected directory entry that caused the trigger to fire.
2. values corresponding to the six parameters to the Trigger constructor.
3. bindAuthorization is a structure containing the login name and password of the user whose action caused the trigger to fire – it is only used for Before triggers (which can only be written by super-users), and only from LTAP servers that have been configured at start up time to deliver this information. This information is provided in those cases to allow query/update rewrite

---

[7] MetaComm's TAS, the Update Manager, only uses Before triggers, and it never returns Proceed.

```
ActionStatus TriggerFields::triggerBeforeAction() {
    Reformat Before/After image into internal format;
    Use reformatted images to create commands to update
        PBX, messaging platform, and LDAP directory;
    If all updates to individual repositories succeed,
        return StopSuccess;
    Otherwise, log error and return StopFailure;
}
```

Fig. 6. TAS Code for the `Before` trigger, `triggerPersonMod` (part of the MetaComm's Update Manager)

by TASs while still preserving the security authorizations of the requesters – the TAS assumes the requester's access rights while executing the rewritten command. Note that sending this information can be turned off in the LTAP configuration file.

The TAS instantiator's task with regard to processing trigger action service requests is to write three member functions of `TriggerFields`: `triggerBefore Action`, `triggerFailAction`, and `triggerAfterAction`. The provided server code parses each incoming service request. If the parsed request has a `triggerWhen` value of `Before`, the server calls the `triggerBeforeAction` member function and returns the `ActionStatus` returned by that call to LTAP. If `triggerWhen` is `After` (`Failed`), the server calls `triggerAfterAction` (`triggerFailAction`) and returns `Proceed` to LTAP after the call completes. Since the value is ignored by LTAP for `After` and `Failed` triggers, the server code picks `Proceed` arbitrarily to send back for the purpose of synchronization with the LTAP gateway. A TAS that handles the trigger `triggerPersonMod` of Fig. 4 can be constructed by defining the `triggerBeforeAction` as in Fig. 6.

5   Experiences

In this section we discuss the rationale for various design decisions as well as future extensions to LTAP based on our experiences in building MetaComm. We describe tricks we developed for exploiting LTAP features, and useful interactions between LTAP and our data integration system, lexpress [28]. We conclude the section with a discussion on how to use the LTAP gateway model to provide advanced database functionality to directories.

5.1   Extensions to LTAP

Persistent Connections and Synchronization: One useful feature that MetaComm provides is synchronization. Synchronization is useful for initially loading data from a device into the global integrated directory, as well as ensuring consistency across devices after catastrophic failures occur. In order to provide the synchronization facility, MetaComm must guarantee that after a synchronization request is processed, the LDAP server, the device being synchronized, and other devices that share data with the LDAP server or device are consistent. Even though synchronization requests can be thought of a sequence of individual updates, the set of updates should be applied in isolation, *i.e.*, other updates should not be allowed concurrently. This required modifications to LTAP:

1. In its original implementation, LTAP only allowed a single update per connection from LTAP to a TAS (*e.g.*, UM), but to differentiate synchronization requests from individual updates, persistent connections were added which allow a sequence of updates.
2. In order to guarantee that synchronization requests are executed in isolation, all updates must be disallowed while a synchronization request is being processed. To support this, a new *quiesce* facility was added to LTAP.

Monitored Fields: When LTAP was originally designed, only the `triggerMonitoredFields` construct was provided – this construct combined the functionality of `triggerMonitoredFields` and `triggerFieldsOfInterest` in the current system. However, one of our users insisted that they be separated to reduce the amount of work needed to process a trigger. For the application in question, a notification that a field had changed was sufficient – for instance, if a person's name and telephone number have changed, the application only needs to know the name of the person but nothing more. Thus, the distinction between monitored fields and fields that must be shipped to a TAS was made.

This distinction was important for MetaComm as well, but for a different reason. MetaComm only needs to know if name and telecomm-related fields are about to change. However, when it must handle the change by translating the update, it may need information about other fields which have not been modified. For instance, if the IAA asks to modify the `sn`, `telephoneNumber`, and `description` field of a `person`, the UM will be alerted because of the first two fields. When it rewrites the LDAP command, it will also need to know the new value for `description`, so that the semantics of the update are preserved.[8] Given that new auxiliary classes can be added to an existing object at any time, it was not practical for MetaComm to explicitly list all the fields that it would like to have shipped, so it asks that all fields belonging to a added/deleted/modified entry be shipped.

TAS Return Values: While developing MetaComm, we discovered the need to support more return values from a trigger action server (*e.g.*, the Update Manager). Besides `Proceed` and `StopFailure`, we also needed `StopSuccess`. The two original values are adequate if the server only needs to approve or reject the operation. However, the MetaComm implementation could be simplified if it updated both the devices and the directory, so that it could deal with failure

---

[8] In addition to the fields mentioned explicitly, it will need other fields like `cn` that will implicitly be changed by the change to `sn`. See Sect. 5.2.

of LDAP commands more easily. Otherwise, if it had to deal with a `Failed` trigger, it would have had to keep around state from the device updates until it was certain that the LDAP command had gone through – leading to garbage collection problems.

5.2   Exploiting LTAP Features

Handling Anomalies and the Need to Rewrite Updates: LDAP commands that come to MetaComm sometimes need to be rewritten. For instance, if Jill Lu's `sn` is changed to `"Lu-Smith"`, then her `cn` should be changed to `"Jill Lu-Smith"`. LDAP does not enforce such dependencies, but a simple operation rewrite by lexpress in the UM can. Also, certain operations on the devices return information that needs to be reflected in the directory. For instance, when an entry for a new mail box is created on the messaging platform, an id is generated by messaging platform. Since this id is needed by the LDAP directory, the messaging platform passes the id to the UM which rewrites the original LDAP command to set the corresponding field in the LDAP directory. The Update Manager then sends the rewritten LDAP command to the LDAP directory directly. If it succeeds, it returns `StopSuccess`. Otherwise, it returns `StopFailure`.

Given that updates needing rewrites will be directly applied by the UM to the LDAP directory, it was simpler to handle all of them this way. The alternative was to have LTAP directly apply the LDAP commands and have a second `Failed` trigger handle update failures. Thus, we would expect trigger writers to typically write either `StopFailure/Proceed` or `StopSuccess/StopFailure` triggers exclusively.

Reduced Numbers of Triggers: When originally designing LTAP, we anticipated that users would set triggers on the type highest in the lattice whose modification was of interest, but that only fields belonging to that type would be sent to the TAS. This turned out not to be practical, since some application need to differentiate fields to be monitored and fields to be included in a trigger notification (see Sect. 5.1). While developing MetaComm, we noticed that we could take advantage of the fact that the only updates we were interested were updates to `person` entries. The update might be to fields belonging to an auxiliary class that was hung off a `person`, but the `objectclass` of the modified entry would include `person` as one of its types. By placing the trigger high in the class hierarchy and taking advantage of LTAP's willingness to allow the triggers to specify any field whatsoever (including those of `person` subclasses) as being of interest and to be monitored, MetaComm was able to get by using only three triggers – one for modify, add, and delete of a `person`. A single *any-update* trigger would handle all the alerting needs for the part of the system described in this paper. The extensions to MetaComm that are ongoing will probably require only one trigger of the *any-update* sort for each telecomm-related class.

The form of all of MetaComm's triggers was extremely similar to that of `triggerPersonMod` in Fig. 4. All the triggers are interested in monitoring the same set of fields and in having all fields sent to the UM. The differences were quite uniform. The second parameter to the `Trigger` constructor is different for each, specifying either add, delete, or modify. Also, the `triggerNeeds` fields specifies `NeedsAfter` (`NeedsBefore`, `NeedsBoth`) for the trigger for adds (deletes, modifies). As a result of this experience, we realized the need to add an *any-update* trigger type in a future version of LTAP.

Using Pre-images: We initially decided to receive pre-images, as well as post-images, in LTAP notifications – to give us access to the keys for updating the devices. The keys for updating the devices are often in other attributes than the distinguished name for the LDAP object. When the key changes, both the old and the new value of the key are needed to update the target device. For instance, a key modify may require moving a user from one PBX to another. In practice, the pre-image was even more useful. The lexpress component uses pre-images during constraint processing to determine if the object being modified resides on the target device. It compares pre-images and post-images to determine what changes have occurred, and only produces updates for target attributes that must change. In addition, lexpress must at times use the values of source attributes that have not changed in creating new values for target attributes. lexpress creates a composite target attribute from multiple source attributes. The composite attribute changes when at least one of its contributing source attributes change. Thus, when creating a new composite attribute value, lexpress requires values for all of the source attributes, whether they have changed or not.

5.3   Offering other Extended LDAP Functionality through Gateways

The use of a gateway to add needed but unstandardized functionality to an LDAP directory (or any repository that speaks a standard protocol) was validated by the experience of LTAP in MetaComm. Such gateways offer portability advantages. For instance, while preparing the demo for InterOp [27], we were able to change directory brands easily – had we relied on proprietary trigger facilities, likely the port would have been much harder. Another advantage of the gateway approach is that it only makes collections of applications that need a given piece of functionality pay for the overheads of the functionality – simpler application suites can go directly to an un-encapsulated directory.

The same gateway technique could also be used for replication or materialized views. Stronger security techniques could also be built into a gateway than are present in the directory, provided one can ensure that there is no direct path to the directory except through the gateway and a handful of trusted servers. It would be attractive to use a similar idea to provide sagas [12]. For example, a simple saga mechanism could be built by creating special messages to the gateway to indicate transaction begin and end and to treat each intervening request as a step in the saga, and by keeping information about what needs to be undone if any operation returns `StopFailure`. Support for more general transaction models would be useful, but this may not be possible when the underlying repositories do not have two-phase commit support.

6 Related Work

Details of the LTAP gateway are given in [24]. In this paper, we focus on the use of LTAP in the MetaComm system – its benefits and extensions/changes that were needed. The MetaComm system and its relation to previous work (*e.g.,* data integration [23,11], sagas [12], data warehouses and views [2,4,29,14]) is described in [10]. Unlike [10], this paper focuses on our experiences in using LTAP to build MetaComm, the features of LTAP that we exploited in novel ways, and our rationale for various changes and extensions to LTAP.

Many directory servers (*e.g.,* Microsoft Active Directory) have no support for triggers, even though plans have been announced to add the functionality. Netscape's directory plug-ins can be used as triggers since they can be associated with directory operations [26]. They have structures similar to LTAP triggers – both associate an action with the time immediately before or after the performance of an LDAP command. Netscape's plug-ins run as function calls in the directory server itself, and even though this has performance advantages, it causes reliability, portability, and availability problems that LTAP does not have [24]. Given that MetaComm needed to be supported across all LDAP servers, LTAP was the only viable choice.

Triggers are a well-known technique for databases. A user specifies an event (or pattern of events) of interest that the system is to monitor, a condition to check (which may always be true), and an action to take. When the event is detected and the condition holds true, the system performs the action. The novelty of our LTAP work is our support for `Failed` triggers, necessary to cope with the weak transactions of LDAP, and the use of a proxy to provide trigger support. Previous techniques for adding triggers detected events either by adding event detection into the core database engine, *e.g.,* [3,34,25,31], and/or by using pre-processors on database code to signal events, *e.g.,* [25,6].

C^2*offein* is a system to add ECA triggers in a CORBA environment [21]. They get their events from wrappers around data sources. Rather than write special data wrappers for each data source, LTAP takes advantage of the standard vocabulary of LDAP and only requires one "wrapper" to support arbitrary LDAP data sources.

Two recent proposals for LDAP triggers [13,32] suggest the use of persistent queries where a query stays active finding adds, deletes, and modifies to the LDAP directory that match the search condition. A persistent query, thus, corresponds to a trigger. This mechanism has scalability problems since it must maintain one open connection per persistent query. In contrast, LTAP requires an open connection only for triggers that are currently being processed. Persistent search, once standardized, could be used to build a tool providing general `After` triggers in a scalable way – in other words, a tool providing a subset of LTAP functionality. Such a tool would eliminate the scalability concerns, since only a single connection to the tool would be required. However, `Before` and `Failed` triggers could not be supported by such a tool, since it would only be alerted after a successful modification had already taken place. Furthermore, such a tool could not provide pre-update values unless it keeps its own copy of the directory. These limitations would make it harder to write an application like MetaComm.

Other work to extend LDAP to take advantage of database technology aims at adding a more expressive, non-procedural query language to LDAP [18]. We share their aim of moving useful techniques from databases into directories, but deal with triggers rather than queries. The architecture of the LTAP library version is very similar to that used to add more flexible list management to LDAP [19]. In their work, queries to LDAP go through a "list location and expansion" client library before being sent to the LDAP directory. In the LTAP library approach, triggers are provided by having LDAP queries and updates go through a library that does trigger processing as well as passing commands on to the LDAP directory.

7   Conclusions and Future Work

Triggers and other basic database functionalities are needed by LDAP servers to support applications such as Directory Enabled Networking. The lack of standard trigger facilities is a significant barrier to progress as users begin using LDAP for tasks like provisioning network services, allocating resources, reporting, managing end-to-end security, and offering mobile users customized features that follow them. Current LDAP servers either lack trigger facilities or provide proprietary triggers with significant limitations. In this paper, we described the use of LTAP, a gateway which provides a portable way to add full-fledged triggers to LDAP servers, in the telecom integration project MetaComm [10]. LTAP's alerting and locking facilities provide valuable glue for combining the directory, a Definity® PBX, and a messaging platform into an integrated package. The resulting platform supports powerful, directory-enabled telecom applications such as the ones described in Sect. 3.3.

We also described our experiences in using LTAP in MetaComm, both where its functionality could be used in unanticipated ways to solve problems, and where it needed to be extended. These experiences are key lessons learned in this paper. They will be valuable to those building similar systems.

The gateway technology appears to be a good choice for providing replication, security, and saga facilities to directories in a portable way. We intend to explore some of these directions in future work.

We would also like to incorporate LTAP into other networking/telecommunications projects to find new enhancements to the system. We are currently extending LTAP's functionality and integrating it with other telecom products. MetaComm technology is currently being transitioned and hardened for commercial use.

Acknowledgments. Thanks to Arun Netravali for suggesting we explore the DEN area. Thanks to Qian Ye who helped build MetaComm and test LTAP. Thanks for valuable comments from Mike Holder, Gavin Michael, Shamim Naqvi, Hector Urroz and G. A. Venkatesh. In addition, we would like to thank our colleagues on the MetaComm project who have not already been mentioned:

Lalit Garg, Julian Orbach, and Luke Tucker. Andy Liao provided us with a Java LDAP parser which allowed us to replace the C LDAP parser from the University of Michigan that we had been using.

References

1. R. Arlein, N. Gehani, and D. Lieuwen: LTAP trigger gateway for LDAP directories. http://ltap.bell-labs.com/LTAPTM.doc
2. J. Blakeley, P. Larson, and F Tompa: Efficiently updating materialized views. In Proc. SIGMOD International Conference on Management of Data, pages 61–71, 1986
3. A. Buchmann, J. Zimmermann, J. Blakeley, and D. Wells: Building an integrated active OODBMS: Requirements, architecture, and design decisions. Proc. International Conference on Data Engineering, pages 117–128, March 1995
4. S. Ceri and J. Widom: Deriving production rules for incremental view maintenance. In Proc. International Conference on Very Large Data Bases, pages 577–589, 1991
5. D. W. Chadwick. Understanding x.500 - the directory, 1994. http://www.salford.ac.uk/its024/Version. Web/Contents.htm
6. S. Chakravarthy, V. Krishnaprasad, E. Anwar, and S. Kim: Composite events for active databases: Semantics, contexts and detection. In Proc. International Conference on Very Large Data Bases, pages 606–617, August 1994
7. S. Cluet, Olga Kapitskaia, and D. Srivastava: Using LDAP directory caches. Proc. Principles of Database Systems, 1999
8. U. Dayal, B. Blaustein, A. Buchmann, U. Chakravarthy, M. Hsu, R. Ladin, D. McCarthy, A. Rosenthal, and S. Sarin: The HiPAC project: Combining active databases and timing constraints. SIGMOD Record, 17(1):51–70, March 1988
9. A. Demers, D. Greene, A. Hauser, W. Irish, J. Larson, S. Shenker, H. Sturgis, D. Swinehart, and D. Terry: Epidemic algorithms for replicated database maintenance. Proc. ACM Symp. on the Principles of Distr. Computing, pages 1–12, August 1987
10. J. Freire, D. Lieuwen, J. Ordille, L. Garg, M. Holder, H. Urroz, G. Michael, J. Orbach, L. Tucker, Q. Ye, and R. Arlein: MetaComm: A meta-directory for telecommunications. Proc. International Conference on Data Engineering, March 2000 (to appear)
11. H. Garcia-Molina, Y. Papakonstantinou, D. Quass, A. Rajaraman, Y. Sagiv, J. D. Ullman, V. Vassalos, and J. Widom: The TSIMMIS approach to mediation: Data models and languages. Journal of Intelligent Information Systems, 8(2):117–132, 1997
12. H. Garcia-Molina and K. Salem: Sagas. Proc. SIGMOD International Conference on Management of Data, 1987
13. G. Good, T. Howes, and R. Weltman: Persistent search: A simple LDAP change notification mechanism. http://www.ietf.org/internet-drafts/draft-ietf-ldapext-psearch-01.txt
14. T. Griffin and L. Libkin: Incremental maintenance of views with duplicates. In Proc. SIGMOD International Conference on Management of Data, pages 328–339, 1995
15. Directory Enabled Networking Ad Hoc Working Group: http://murchiso.com/den/
16. T. Howes and M. Smith: LDAP: Programming Directory-enabled Applications with Lightweight Directory Access Protocol. Macmillan Technical Publishing, 1997

17. Innosoft. Innosoft's LDAP world implementation survey. http://www.critical-angle.com/dir/lisurvey.html

18. H. Jagadish, L. Lakshmanan, T. Milo, D. Srivastava, and D. Vista: Querying network directories. Proc. SIGMOD International Conference on Management of Data, 1999

19. H. V. Jagadish, M. Jones, D. Srivastava, and D. Vista: Flexible list management in a directory. Proc. Conference on Information and Knowledge Management, November 1998

20. A. Keller. *Updating Relational Databases through Views*. PhD thesis, Stanford University, 1995

21. A. Koschel and R. Kramer: Configurable event triggered services for CORBA-based systems. Proc. International Enterprise Distributed Object Computing Workshop, November 1998

22. http://www.ietf.org/html.charters/ldapext-charter.html

23. A. Levy, A. Rajaraman, and J. Ordille: Querying heterogeneous information sources using source descriptions. Proc. International Conference on Very Large Data Bases, pages 251–262, 1996

24. D. Lieuwen, R. Arlein, and N. Gehani: The LTAP trigger gateway for LDAP directories. Software Practice and Experience (to appear)

25. D. Lieuwen, N. Gehani, and R. Arlein: The Ode active database: Trigger semantics and implementation. Proc. International Conference on Data Engineering, pages 412–420, February–March 1996

26. http://developer.netscape.com/docs/manuals/directory/plugin/index.htm

27. Lucent Press Release. http://www.lucent.com/press/ 0599/990503.nsc.html

28. J. Ordille. Mapping updates for heterogeneous data repositories. Technical report, Bell Laboratories - Lucent Technologies, 1999

29. X. Qian and G. Wiederhold: Incremental recomputation of active relational expressions. IEEE Trans. on Knowledge and Data Engineering, **3**(3):337–341, September 1991

30. L. Seligman and L. Kerschberg: A mediator for approximate consistency: Supporting "good enough" materialized views. Journal of Intelligent Inf. Sys., **8** (1997) 203–225

31. M. Stonebraker, E. Hanson, and C. Hong: The design of the Postgres rules system. Proc. International Conference on Data Engineering, pages 365–374, 1987

32. M. Wahl: LDAPv3 triggered search control. ftp://ftp.isi.edu/internet-drafts/draft-ietf-ldapext-trigger-01.txt

33. M. Wahl, T. Howes, and S. Kille: Lightweight Directory Access Protocol (v3), December 1997. http://www3.innosoft.com/ldapworld/rfc2251.txt

34. J. Widom, R. J. Cochrane, , and B. Lindsay: Implementing set-oriented production rules as an extension to Starburst. Proc. International Conference on Very Large Data Bases, pages 275–285, September 1991

Requirements Analysis of Distribution in Databases for Telecommunications

Juha Taina and Kimmo Raatikainen

Department of Computer Science, P.O. Box 26,
FIN-00014 University of Helsinki, Finland
{Juha.Taina,Kimmo.Raatikainen}@cs.Helsinki.FI,
http://www.cs.Helsinki.fi/{Juha.Taina,Kimmo.Raatikainen}

Abstract. Current and future telecommunication systems will rely heavily on some key IT baseline technologies including databases. We present our analysis of distribution requirements in databases for telecommunications. We discuss those needs in GSM and IN CS-2. Our requirements analysis indicates that the following are important issues in databases for telecommunications: single-node read and temporal transactions, clustered read and temporal transactions, distributed low-priority read transactions, distributed low and middle level management transactions, support for partitioning and scalability, dynamic groups, replication control, deadlines for local transactions, and limited global serialization control.

1 Introduction

Databases in telecommunications have gained a growing role in the last few years. The trend started at the time when digitalization started to supersede the analogous architectures and since then the role has expanded rapidly. Today, modern telecommunication networks could not operate without databases.

Although databases in telecommunications have been around for over a decade, little work has been done to identify the requirements for them. In our projects DARFIN (Database ARchitecture For Intelligent Networks) and RODAIN (Real-time Object-oriented Database Architecture for Intelligent Networks) we have identified the general requirements of databases in Intelligent Networks, mostly in the view of Service Data Function (SDF). [17] Other work that cover the area are [12] and [3]. All these works concentrate on the database requirements in Intelligent Networks.

The results of the studies are clear. Real-time data access and fault tolerance are the key issues in IN databases. Data requests must be served quickly and reliably in all workloads. This is common to all service-related database requests. On the other hand, management requests can be served slower, yet the fault-tolerance level must be very high. The same result can be expanded to other service-oriented telecommunications areas, such as regular fixed telecommunication networks and mobile telecommunication networks. Wherever service-related requests are needed, the database management system must respond quickly and it must be continuously available. Whenever management requests are needed,

W. Jonker (Ed.): Databases in Telecommunications, LNCS 1819, pp. 74–89, 2000.

the database management system must be extremely reliable and logically consistent.

All the results above summarize a non-distributed database management system. Such a system is a single database node that can serve requests from several applications, some of which might be geographically far from the database. However, the telecommunications networks are by default distributed. Every routed call has two distributed switches; one on the caller's side and one on the called side. The switches can be the same but it is not necessary. Hence, a natural question is whether telecommunications databases should also be distributed and what kind of distribution is needed.

In this paper we cover the area of distribution requirements for databases in telecommunications. We examine the requirements in GSM and in IN CS-1 and CS-2.

2   Data Distribution in GSM

The GSM database requirements form a good case of current telecommunications database needs. The architecture is well specified and working, and a good summary can be collected from several sources [11,14,15]. It gives a good overview of the network requirements, including the database requirements.

The GSM architecture can be divided into three subsystems: the Base Station Subsystem (BSS), the Network and Switching Subsystem (NSS), and the Operation Sub-System (OSS). The three subsystems have different database needs starting from no requirements and ending in strong real-time requirements.

The BSS does not have any database requirements. If BSS architecture uses databases, the databases are embedded in BSS elements and are vendor-specific. The BSS architecture does not affect the requirements.

The NSS has four major database management systems that are listed in the GSM standards: the Home Location Register (HLR) that maintains service related subscriber information, the Visitor Location Register (VLR) that maintains location information about GSM subscribers, the Equipment Identity Register (EIR) that maintains information about GSM subscriber equipments, and the Authentication Center (AC) that is used in security procedures.

HLR is the main database management system in a GSM network. It is an active component of the system and controls call and service related information. The HLR must offer real-time database services. Its main functions include such time-critical tasks as finding subscriber information and informing about the current location of a subscriber. Usually the architecture of HLR is not distributed. A HLR can handle hundreds of thousands of subscribers. If more capacity is needed, several HLRs can be executed simultaneously. This can be achieved since subscriber data is hierarchical.

VLR has a much smaller role than HLR. It is responsible of maintaining exact location information of subscribers that are in the area of all of its Mobile Switching Centers. It communicates with the HLRs when calls are routed or charging information must be retrieved. Hence, it can be considered as a specialized database node in a distributed real-time database. It has real-time and distribution requirements. However, the distribution requirements are not strict. Updates occur only in one VLR, and the only distribution is into known HLRs. Since the information location is always known and a single HLR, distribution basically degenerates into message exchange.

The last two databases, EIR and AC have very specialized roles. The EIR has a management role and it does not have real-time requirements. It has distributed requirements since it can communicate with other EIRs. Nevertheless the nature of the EIR is a regular distributed database. The AC needs to serve real-time requests, but it does not need distribution. It is often connected to the HLR, or a single database management system offers both HLR and AC services.

From the small analysis above we collect the HLR/VLR relationship in Figure 1. The relationship does not include EIR and AC. Both could be added, although EIR is basically distinct from other databases.

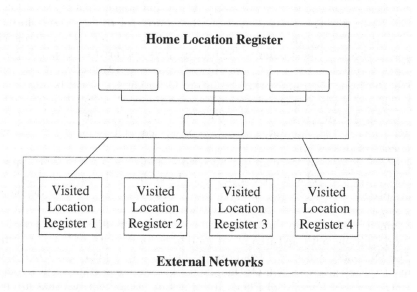

Fig. 1. GSM HLR/VLR Relationship

The HLR architecture is best described as a parallel database architecture where several database nodes are connected with a fast bus to form a logically uniform database management system. It exchanges information and messages with a set of VLRs, some of which are in external networks. The HLR has distributed requests but they degenerate into message passing between the HLR and one or more VLRs. Data is not replicated and data location information is well known. All real-time reads occur only in a single node, since the needed information is subscriber information which is hierarchical.

Finally, the OSS can have several database management systems although the GSM standard does not specify them. The management databases need not be real-time databases. Although requests must be served rapidly, it is enough that they are served with high throughput databases rather than real-time databases with deadlines.

On the other hand, the distribution level can be very high in OSS. Since the whole network must be managed and the components are geographically distributed, distributed database architectures can also be useful (although not necessary).

3   Data Distribution in Intelligent Networks

The basic idea of Intelligent Networks (IN) is to move service logic from switches to special IN function logics called Service control functions (SCF). With SCFs it is possible to let telephone switches be relatively static when new service logic programs are used in SCFs to implement special services. Hence IN architecture can be used for special service creation, testing, maintenance, and execution.

The basic standard that defines the framework of other IN standards in ITU-T is the Q.1200-series. It defines a framework for ITU-T IN standards with possible revisions. The standards themselves are structured to Capability sets. The first Capability Set (CS-1) was introduced in Q.1210-series at the same time as the basic IN framework. The future Capability sets will be introduced in Q.12n0-series where n is the capability set number. This Section is based on IN CS-1 in Q.1210-series. The capability sets are compatible backward to the previous ones so that the implementation of services can be progressed through a sequence of phases [4].

The capabilities in CS-1 are intended to support IN services that apply to only one party in a call and is independent at both the service and topology levels to any other call parties [7]. In other words, the services are only dependent on a single call party and the same service may be used again in a different environment with possibly different parameters. This definition restricts the CS-1 service definitions but it in turn gives a good basis for the future IN architectures.

The IN conceptual model has been divided into four planes that define different views to the IN model. Of the planes the most important to distribution is the Distributed Functional Plane. It defines functional entities that are used for IN service implementation. The entities are visible only on the Distributed Functional Plane.

A functional entity is a group of functions in a single location and a subset of the total set of functions that is required to provide a service. Every functional entity is responsible for certain actions. Together they form the IN service logic. Thus, the Distributed Functional Plane defines the logical functional model of the IN network. The functional entities that CS-1 defines are: the Call Control Agent Function (CCAF), the Call Control Function (CCF), the Service Creation Environment Function (SCEF), the Service Data Function (SDF), the Service Management Access Function (SMAF), the Service Management Function (SMF), the Specialized Resource Function (SRF), and the Service Switching Function (SSF).

Starting from the IN CS-1 description, we are ready to analyze IN database management systems in further detail. This analysis is based on both IN CS-1 and IN CS-2. IN CS-2 is included because the current draft version is already publicly available. It is probable that the drafts are stable and will have only minor modifications in the future. The IN CS-2 architecture specifies IN functional entities more accurately than IN CS-1.

Databases have a dominant role in IN CS-2 Distributed Functional Plane (DFP). Not only the Service Data Function has data management but also several other functional entities have either database managers or services that need database management (Figure 2). In the CS-2 architecture, CCF, SSF, SRF, SDF, and SCF have data managers. The SMF architecture is left for further study in CS-2. Since management needs to keep track of managed data, a database management system is needed in the SMF architecture.

Most of the data managers are embedded in functional entities. Only data managers in SDF and perhaps SMF are global in the sense that they can accept requests from other functional entities. The managers of other entities service requests from inside the entity.

3.1 Service Switching and Call Control Functions

The functional model of SSF/CCF/CUSF is described in [10]. The most interesting functional entities in the model are two data managers; the IN local resource data manager and the Basic call resource data manager. Neither of the managers has been described in CS-1, except that they are assumed to exist [8]. The CS-2 draft only states that CCF manages basic call resource data which consists of call references [10]. Hence, it is difficult to say much about the requirements for the data managers. We assume that the IN local resource data manager is responsible for managing data that is related to triggering IN services. Such data has a simple structure. A manager must be tailored for fast data access with a simple key. As for the Basic Call Resource Data Manager, we assume that it is responsible for managing dynamic call connection data. The managed data is temporal. Static data may be present if routing information is kept in the database.

3.2 Special Resource Function

The Special Resource Function (SRF) offers specialized resources for IN services that are beyond Service Control Function services. For instance, SRF is responsible for voice recognition, conference bridges, etc. As such, it is a very specialized functional entity with its own database requirements.

The most interesting entity of the CS-2 SRF architecture is the Data Part (DP). The ITU-T Q.1224 draft states that the DP is composed of a Database manager and a database that contains recorded voice, sound, image, text, etc [10].

The SRF Data Part is a multimedia database. As such, it has to offer fast data access to complex data. The database is defined as a server for internal data

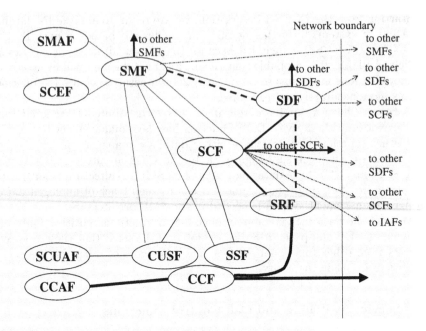

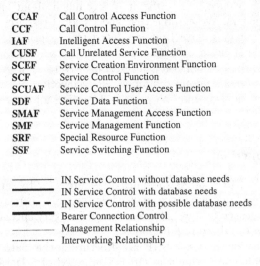

CCAF	Call Control Access Function
CCF	Call Control Function
IAF	Intelligent Access Function
CUSF	Call Unrelated Service Function
SCEF	Service Creation Environment Function
SCF	Service Control Function
SCUAF	Service Control User Access Function
SDF	Service Data Function
SMAF	Service Management Access Function
SMF	Service Management Function
SRF	Special Resource Function
SSF	Service Switching Function

——————— IN Service Control without database needs
▬▬▬▬▬ IN Service Control with database needs
– – – – IN Service Control with possible database needs
▬▬▬▬▬ Bearer Connection Control
——————— Management Relationship
·················· Interworking Relationship

Fig. 2. Q.1224/IN CS-2 Distributed Functional Plane Architecture

requests. The SRF can also be a client for an SDF since it can accept service logic scripts from SCF that can include SDF data requests.

The SRF management has been left for further study in CS-2. The recommendations state that SMF is responsible for the management of the service specialized resources, such as User Interaction scripts, resource functions, and data. It is also possible that service subscribers can manage their private data in SRF. Together these request types imply that the SRF must offer a database interface for external requests.

3.3   Service Control Function

A similar structure as in SSF/CCF/CUSF is present in the CS-2 SCF model (Figure 3 and Figure 4). The architecture model has two data managers; the Resource Manager and the SCF Data Access Manager. The Resource Manager controls the local SCF resources and provides access to network resources in support of service logic program execution. The former resources are needed only on SCF execution and can be considered an integrated member of the architecture. The latter resources, while referencing global network related data, are accessed via the other data manager, SCF Data Access Manager. Thus, the Resource Manager sets requirements only to the SCF Data Access Manager and not directly to an IN database architecture.

The SCF Data Access Manager offers the functions needed to provide storage, management, and access of shared and persistent information in the SCF. The data access manager also provides the functions needed to access remote information in SDFs [10]. This implies that the data access manager is both a client to one or more SDFs and a database manager for SCF local data.

The SCF Data Access Manager manages two types of data; the Service Data Object Directory and the IN Network-Wide Resource Data. The Service Data Object Directory provides means to address the appropriate SCF for access to a specific data object [10]. This implies that the data object directory is used for SCF interaction instead of SDF data requests. However, the recommendations also state that the Service Logic Element Manager, which is responsible for service logic programs, interacts with the data access manager to access service data objects in SDFs. The SCF Data Access Manager uses the Service Data Object Directory to locate service data objects from SDFs [10]. This implies that the Service Data Object Directory is used for SDF access. We assume that the SCF Data Access Manager is responsible for all remote data access regardless of its location. It can access both SCFs and SDFs.

The other data element, IN network-wide resource data, is defined as a data base for information about location and capabilities of resources in the network that can be accessed by the executed service logic programs [10]. Furthermore, the recommendation states that the addressed functional entity is usually a service resource function SRF. Hence, the IN network-wide resource data is location information about useful service resources, mostly SRFs. An SCF requests special services from an SRF which executes the request. The SCF may forward a service script for execution or request a single operation [10].

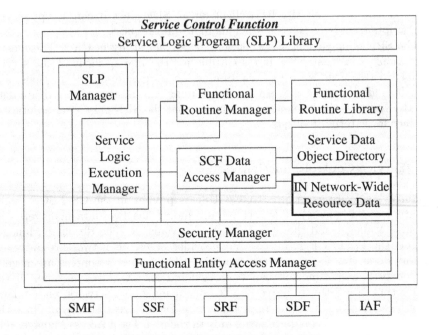

Fig. 3. Q.1214/IN CS-1 Service Control Function Architecture

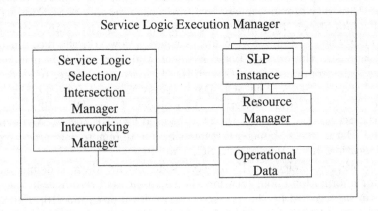

Fig. 4. Q.1214/IN CS-1 Service Logic Execution Manager Architecture

Our conclusion is that an SCF is a database client for SDFs and SCFs, and a service client for SRFs. The recommendations do not state what kind of data is stored in SCFs, but it is probably directly related to service execution. All global service-related data should reside in SDFs.

Next to the dynamic data managers, two data managers are defined in SCF: the Service Logic Program Manager (SLP Manager) and the Functional Routine Manager. The SLP Manager manages service logic programs that constitute IN services. The Functional Routine Manager manages operations that implements SCF functions. The managed data of these two managers is static and local to the SCF.

Although it is not explicitly stated, we assume that SMF is responsible for all SCF service management. All update requests to SCF static data arrive from SMF. Only updates for dynamic data may arrive from other SCFs.

3.4   Service Data Function

The CS-2 SDF architecture model is depicted in Figure 5. The entity is responsible for real-time access and maintenance of global network related data. The ITU-T recommendations state that the SDF contains and manages data which is related to service logic processing programs and accessed in the execution of the programs [10]. This definition limits SDF for a server of requests from SCFs.

According to the CS-2 recommendations, the SDF contains customer and network data for real time access by the SCF. Its tasks include secured data acquisition and management of data, distributed requests with data location transparency to the requester, inter-networking security, optimal distributed data access depending on the network traffic, data replication and access rights for replicated data, authentication and access control for secure access to service data, data support for security services, and fault-tolerance [10]. These tasks are typical for a distributed database management system.

The SDF data types are not as clearly listed. The recommendations state that SDF data types are authenticate data for user authentication and access rights (PIN-codes etc.), private operational data for SDF administration, and service data for the provision of a service. In other words, SDF data can contain anything. A detailed study of database usage in CS-1 services and service features can be found in [16]. There are five basic types of database operations that are needed in CS-1:

1. *Retrieval of structured objects from persistent subscriber data objects.* Some of the retrievals trigger a later update.
2. *User management actions that modify persistent subscriber data objects.* These actions must be protected against unauthorized use. Some updates may also have to be verified against other criteria.
3. *Verification of Personal Identification Number (PIN).* For security reasons the PIN verification must be done in the database system when services are globalized.
4. *Writing sequential log records.*

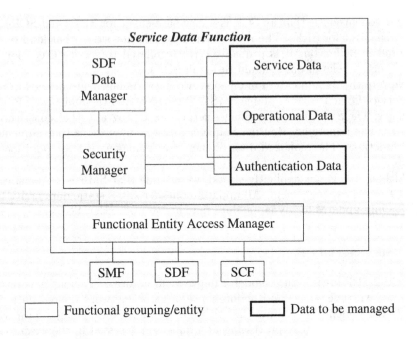

Fig. 5. Q.1214/IN CS-1 Service Data Function Architecture

5. *Mass calling and Televoting.* The operations need high-volume small updates that must be serialized. A special aspect of these updates is that they can be done in large blocks.

The most common ones of these operations are retrieving structured objects and writing sequential log records. Security operations, such as PIN verification, are also common. Management updates are not as common as the previous types. And finally, mass calling and televoting occur only on special occasions. They may also need SCFs that are specifically tailored for that use. As for IN CS-2, the third item (PIN verification) must be expanded to support general authentication and security services, otherwise the list holds true in CS-2 as well.

The relationships between functional entities in IN DFP define how the entities communicate with each other. The Intelligent Network Application Protocol (INAP) that is required to support Functional Entity communication in CS-1 is defined in [9]. The INAP will support any mapping of functional entities into physical network entities. Thus the protocol is defined for maximal distribution, e.g. that every network entity consists of a single functional entity.

The INAP is a collection of Application Service Elements (ASE) that provide the necessary communication capabilities between functional entities. The ASE that is defined in CS-1 for SCF-SDF communication is called User Data Manipulation. It consists of two operations; Query and UpdateData. In [1] it is claimed that the ASE is oversimplified. It has two problems; service-dependent semantics of operation parameters and authentication. The service-dependent semantics imply that the actions an SDF must perform are also service-dependent. The result of this is that every SDF must be tailored for each service. Authentication problems occur when an SDF is accessed from an external network. The SDF must perform the authentication of users and provide access control. In the ASE specification the only way to carry out authentication is to use attribute value comparison. The ASE does not define when the comparison should take place. In CS-2, the INAP interface exists for backward compatibility reasons. However, the recommended new SDF interface is based on X.500 Directory Access Protocol (DAP). The new protocol fixes the problems with ASE and tailored SDFs.

The SDF data management is not addressed in CS-2. Since SDF offers database management services, it should manage its own data. Write requests arrive mostly from SCFs and SDFs when the service data is changing. In case of larger updates, such as adding new services and customers, the update request should arrive from the SMF. Hence, SDF at least serves all types of SCF and SDF requests, and large SMF update requests.

3.5   Service Management Function

The SMF has been left for further study in CS-2. Here we assume that SMF will be based on the Telecommunications Management Network (TMN) since one of the possible long term IN architecture plans are to be compatible with Telecommunications Information Networking Architecture (TINA). TMN is widely used in telecommunications and one of the goals of the TINA consortium is to have a common foundation between IN and TMN [2].

TMN [M.3010] is a generic architecture to be used for all kinds of management services. It is based on the principles of the OSI Management [6]. The fundamental idea in the OSI Management is that the knowledge representing the information used for management is separated from the functional modules performing the management actions. OSI Management is based on interactions between management applications that can take the roles of manager and agent. The interactions that take place are abstracted in terms of management operations and notifications. Management activities are effected through the manipulation of managed objects (MOs).

An agent manages the MOs within its local system environment. It performs management operations on MOs as a consequence of management operations issued by a manager. An agent may also forward notifications emitted by MOs to a manager. The agent maintains a part of the Management Information Tree (MIT) that contains instances of MOs organized as a hierarchical database tree. In brief, the principles of OSI Management (and TMN) require that the database system contains the functions of an OSI Management Agent.

If IN service management is based on TMN, the managed objects are all IN elements, including IN Distributed Functional Plane entities. The manager of a functional entity has knowledge of the element's management needs. The SMF needs database services for maintaining this information. TMN architecture can be used for all network management, not only for IN element management. The more management is left for TMN, the more database services are needed.

TMN management and offered database services are equivalent to SDF database management and offered services. Thus, a database node that offers both TMN and SDF services is both a manager and a managed network element. This implies that the TMN/IN SDFs have at least two interfaces; one for IN that is based on INAP or X.500, and one for TMN that is based on X.700. This in turn causes overhead when data is translated between models.

A problem arises when data is shared between IN and TMN. If each interface would address different data there would not be a problem. IN data could be managed locally without interference with IN services. However, this is clearly not the case. For instance, user profile and routing information must be available both to TMN and IN. Either IN and TMN databases must be separated from each other and some data is duplicated to both IN databases and TMN databases, or the same data must be available in a database with different data model mappings to IN and TMN.

3.6 Distributed Database Requirements in CS-2

As we have seen, the CS-2 DFP has several database management systems. We classify them into two groups; embedded systems and external systems. An embedded database management system is an integral part of the functional entity where it is defined. An external database management system has clients at other functional entities, or it offers distributed database operations with other external systems.

The functional entities in CS-2 that have internal database management systems are CCF, SSF, and SRF. A CCF has a Basic Call Resource Data manager, an SSF has an IN Local Resource Data manager, and an SRF has a Data Part. These database managers are so deeply integrated into the appropriate elements that their requirements are beyond the scope of a general analysis. In short, the SSF/CCF requires fast data access from databases because they are related to connecting calls and triggering IN services. The SRF database is a multimedia database. The requirements of such a system are beyond this paper, but a comprehensive analysis of multimedia databases can be found in [13].

The IN Distributed Functional Plane has two external database management systems; the Service Data Function, which has external clients, and the Service Management Function which must have distributed database features for management. In this report we concentrate on the distribution requirements of IN Service Data Function. Although the SMF database architecture is distributed, it is not very important since it is based on the regular database management systems. Moreover, not much is known about it yet. A detailed study of expected SMF architecture and database requirements can be found in [18].

Database Interfaces and Request Types. An SDF may interact with SCFs, SDFs, and SMFs. Also in CS-2, it is mentioned that an SRF may request data from an SDF, for instance when a customer wants to use a different language in a message than the default one [10]. All these requests are local in a network. In addition, an SDF has to answer requests from SDFs and SCFs in external networks, and requests that come from monitor terminals. Several SDFs can form a distributed database with several entries.

In CS-1, the SDF interface is CS-1 INAP. In CS-2, the defined SDF interface is X.500 Directory Access Protocol. Both of the interfaces must be supported in an SDF database. Next to these, a high-level query interface is needed for requests from monitor terminals.

Hence three interface types are needed:

1. An interface for X.500 DAP requests, which arrive from other functional elements. These are the main source of requests for an SDF. The requests may be both reads and updates. Even if the SMF model is based on TMN and X.700 CMIP, the SDF database is updated via the X.500 interface.
2. An interface for now obsolete CS-1 INAP requests, which must be supported for backward compatibility. These requests should be relatively rare since all implementations should use DAP. Nevertheless, it must be supported.
3. An interface for high-level query language requests that arrive from monitoring staff. These requests are database management requests that arrive directly from a terminal. This interface is needed for management reasons.

The listed three interface types are sufficient for CS-2. However, in the future an interface for Open Distributed Processing (ODP) channel is also needed if the long term IN architecture plan is to be compatible with TINA. The TINA architecture is based on the ODP models and interfaces.

As the name states, the IN Distributed Functional Plane is distributed. This definition encloses both the fact that IN DFP has several functional elements and the fact that the plane can have several instances of the same functional element.

Although IN DFP defines a distributed model, it does not force the architectures to be distributed. In principle, a monolithic SDF could serve all IN data requests in a small network. Such a solution is not practical since the database can easily become a bottleneck. The approach can be used if the workload can be analyzed and the network is reliable enough.

If the IN service database architecture is distributed, a logical distribution level is to define each SDF as a node in a distributed database. The CS-2 model supports this approach well, since an SDF can interact with other SDFs in the same or in an external network. The X.500 data model supports distribution since it is a distributed directory standard. The recommendations state that SDF architecture must support location transparency [10]

If SDF is distributed, data replication also becomes an issue. In principle distribution does not require replication. In practice it does, since otherwise often referenced data has to be requested from an external source. It is better to replicate data closer to the location of the requester.

As a brief summary, the most important real-time distributed database candidate in IN is the SDF data management. Since it is based on X.500, its requirements can be summarized in X.500 distribution requirements.

4   Summary of Database Requirements

In this section we briefly summarize the requirements derived from the previous descriptions.

Single-node read and temporal transactions: When short real-time transactions are needed, the network is the strongest candidate for a bottleneck. All the listed cases showed the same trend; it is possible to restrict most short real-time transactions into a single node. The system should be designed to support this as much as possible.

The same is true of temporal transactions. Temporal data is rewritten regularly. If the writing transaction is distributed, it is no longer possible to give any guarantee that the transaction is finished in time. If this happens, the written temporal data can become inconsistent and affects the whole database consistency.

Clustered read and temporal transactions: The lower priority real-time transactions can be distributed but clustered. With good clustering it is possible to minimize the drawbacks of distribution. However, clustering is not a suitable answer when transactions are very short or when their priority is high. In such situations the real-time transactions should be local to a single node.

Distributed low-priority reads: Lowest priority read transactions can be widely distributed. Such transactions are usually low-priority queries that are not critical to the system. Their concurrency control level may be relaxed.

Distributed low and middle level management transactions: Finally, management transactions can always be distributed. They are not real-time transactions, which implies that they do not need a real-time environment. It is important that such transactions can serialize both with each other and with writing real-time transactions. It is not necessary to serialize with reading real-time transactions since the reading transactions can relax serialization. However, the relaxation state depends on the applications and accessed data.

Partitioning and scalability support: As can be seen from the examples, hierarchies dominate telecommunications databases. This is a positive aspect since it eases the partitioning problems and hence offers new alternatives for scalability. Telecommunications databases have large volumes of data and they expand rapidly. Hence, data partitioning and scalability can be useful in data overloading.

Dynamic groups: Dynamic groups occur frequently in telecommunications databases. Examples can be seen in IN, call control, and management. They can be supported in a real-time environment as long as general reference information of the groups formed is available. With such information it is possible to re-organize data or use replication to minimize real-time distributed transactions.

Replication control: Replication has been mentioned in most of the previous aspects. Hence, good replication control is a necessity in a telecommunications database. Replication should be invisible to the applications. Updating replicated data need not serialize, although this is a data dependent feature.

Local transaction deadlines: Real-time transactions are useful in telecommunications, both distributed and local. But in case of distributed real-time transactions, in most cases it is sufficient to use only local deadlines instead of global ones. Local deadlines benefit the applications as much as global deadlines while not needing a global time coordinator.

Limited global serialization control: Global serialization is needed in management, and partially needed in real-time distributed databases. A sufficient global serialization policy allows the serialization to be relaxed according to data and transaction rules. In telecommunications, a sufficient relaxation is to allow replicated data to be inconsistent in short periods of time while maintaining serialization for network control and customer management.

5   Conclusions

When telecommunications databases attain new roles in the networks, new data requirements also arise. One such requirement is to combine real-time and distribution. In this paper we have represented various database cases for telecommunications and their requirements. Our conclusion of the cases is that it is possible to build a specialized real-time distributed database management system for telecommunications. Such a system must support low and middle priority distributed transactions, replication, scalability, partitioning, dynamic data grouping, and data hierarchies. Fast real-time service transactions need not serialize. Instead, replication can be used to minimize real-time distributed transactions.

Next to the requirements for the database management system, a list of interesting features can be formed from the cases. Roaming objects are useful when the database needs automatic balancing; either when new nodes are added or when the data schema is reorganized. However, they do not fit well into cases where objects have a large amount of relationships. Another area that is useful and necessary is parallel calculation capacity. If the architecture is designed to allow database nodes to inter-operate in their idle time, they can be used for application-specific parallel calculations.

With these requirements a distributed architecture conclusion is interesting. It is not clear if a real-time distributed database management system is needed for telecommunications. Although telecommunications cases have specific real-time and distribution needs, they can however be relaxed. The resulting database management system—while still distributed-is not a real distributed database. Without global concurrency control and global deadlines the database nodes are more like loosely connected autonomous databases. This is a common trend in telecommunications, for instance in GSM HLR-VLR relationships and in IN SDF-SDF relationships. The only exception is in management, where distributed transactions also need concurrency control. Perhaps the most interesting problem, then, in telecommunications databases is how to allow slow distributed management transactions to coexist with the fast real-time service transactions.

References

1. Chatras, B and Gallant, F: Protocols for Remote Data Management in Intelligent Networks CS-1. Intelligent Network '94 Workshop Record, IEEE, 1994.
2. Demounem, L and Zuidweg, H: On the Coexistence of IN and TINA. Proc. TINA'95 Conf., 1995, pp. 131–148.
3. Eurescom: Fact-finding Study on Requirements on Databases for Telecom Services. Eurescom, 1993.
4. Garrahan, J, Russo, P, Kitami, K and Kung, R: Intelligent Network Overview. IEEE Communications Magazine 31, 3 (Mar. 1993), pp. 30–36.
5. ITU-T: Principles for a Telecommunications Management Network. Recommendation M.3010, 1992.
6. ITU-T: Data Communication Networks—Management Framework for Open Systems Interconnection (OSI) for CCITT Applications. Recommendation X.700, 1992.
7. ITU-T: Introduction to Intelligent Network Capability Set 1. Recommendation Q.1211, 1993.
8. ITU-T: Distributed Functional Plane for Intelligent Network CS-1. Recommendation Q.1214, 1993.
9. ITU-T: Interface Recommendation for Intelligent Network CS-1. Recommendation Q.1218, 1993.
10. ITU-T: Distributed Functional Plane for Intelligent Network CS-2. Recommendation (draft) Q.1224, 1996.
11. Jabbari, B: ntelligent Network Concepts in Mobile Communication. IEEE Communications Magazine 30, 2 (Feb. 1992), pp. 64–69.
12. Kerboul, R, Pageot, J-M and Robin, V: Database Requirements for Intelligent Neywork: How to Customize Mechanisms to Implement Policies. Proc. TINA'93 Workshop, Vol. 2, pp. 35–46.
13. Khoshafian, S and Baker, A: MultiMedia and Imaging Databases. Morgan Kaufmann, 1996.
14. Mouly, M and Pautet, M-B: The GSM System for Mobile Communications. Mouly & Pautet, 1992.
15. Pollini, G, Meier-Hellstern, K and Goodman, D: Signaling Traffic Volume Generated by Mobile and Personal Communications. IEEE Communications Magazine 33, 6 (June 1995), pp. 60–65.
16. Raatikainen, K: Information Aspects of Services and Service Features in Intelligent Network Capability Set 1. University of Helsinki, Department of Computer Science, 1994.
17. Raatikainen, K: Database Access in Intelligent Networks. In Intelligent Networks, J Harju, T Karttunen, and O Martikainen (eds), Chapman & Hall, 1995, pp. 173–193.
18. Taina, J: Database Architecture for Intelligent Networks. University of Helsinki, Department of Computer Science, 1997.

Database Requirement Analysis for a Third Generation Mobile Telecom System

Mikael Ronström

Ericsson Utveckling AB University of Linköping

Abstract. The development of ever faster processors have made it possible to adapt database technology in controlling the telecom network. The requirements of telecom applications are however not generally known in the database community. This paper analyses the requirement on a database to handle services and locations for millions of mobile telecom users. The paper is based on a thorough analysis of network protocols in a pre-standardisation project of UMTS, a third generation mobile telecom system.

The requirements, even on databases supporting million users, are within reach for clusters of a dozen computers of rather modest performance. This paper helps database designers understand the specific requirements that large telecom applications put on databases with regard to performance, response times and availability.

1 Mobile Telecommunications

The first Mobile Telecommunication systems were developed a long time ago. The first systems that generated a mass market were, however, the analog systems. These systems are generally called first generation systems. Examples of these are NMT (Nordic standard), TACS (English standard) and AMPS (American standard). These still exist and in some countries still form a substantial part of the market.

The second generation system development was started when the first generation system was beginning to catch a mass market. Second generation systems were digital and made better use of frequencies and also tried to achieve a better sound quality. The better use of frequencies and also the smaller size of the cells made it possible to sell second generation systems to an even larger market. This market is still expanding very quickly. There are three major systems in the second generation. These are GSM (European standard), D-AMPS (American standard) and PDC (Japanese standard).

In parallel with the development of second generation systems, new systems were also developed for use in houses and in urban areas and in the vicinity of a person's home. A large number of such systems were developed; the major systems that survived in this market are DECT (European standard) and PHS (Japanese standard).

Already in 1987 a project was defined in the RACE-program (Research on Advanced Communications) in EU to define UMTS (Universal Mobile Telecommunications Systems). The aim of this project was to study third generation

W. Jonker (Ed.): Databases in Telecommunications, LNCS 1819, pp. 90–105, 2000.

mobile telecommunication networks. An essential part of this study was to look at integration of the B-ISDN network and the mobile network. In the RACE Il-program (1992-1995) a follow-up project was performed.

In this phase the reality of an exploding second generation system (e.g. GSM) changed the surrounding world. In 1993-1995 also WWW took giant steps into defining the next generation broadband applications. Therefore third generation mobile systems must be based on the second generation mobile systems and the network must be able to interwork with Internet. Standardisation of UMTS is proceeding with the aim of having systems available in 2001–2002.

The explosion of the second generation systems and the development of DECT and PHS which has also attracted a large market have driven the development of generation 2.5. In the USA this development led to PCS that was started a couple of years ago (including IS-95). In this generation the existing systems are further developed for higher bandwidth, better use of radio resources by even more efficient use of frequency and even smaller cells. Packet switched data is an important issue in this generation.

In this paper Mobile Telecommunications are studied from a future perspective with as many users as in the current fixed network and also lower prices and therefore also much higher usage than in current systems. The telecom databases found in this application are used to keep track of users and mobile telephones. It also provides a platform for many network based services; it will also provide an interface to intelligent terminals and thereby make it possible to implement terminal-based services. The databases will store information on these services, location information and directory information. This information is rather compact and possible to store in main memory.

Most procedures require rapid responses; in [2] a delay (without queueing) of 1-10 ms (dependent on query type) in telecom databases gave a delay of 0.6 s in the fixed network for call set-up in simulations (on top of this there is a delay in the mobile access network). Therefore telecom database response time must be in the order of 5-50 ms in a loaded system.

Reliability requirements on telecom switches is generally that they should not be down more than a few minutes per year. Requirements on telecom databases supporting millions of users will be at least as tough and most likely even tougher. Thus a goal of less than 60 seconds downtime per year is needed.

Network studies were performed in this work and were based on an estimation that 50% of all the telecom users will be mobile users. These users will be charged as for calls from fixed phones and therefore their usage is similar to the fixed network. Traffic models were derived from mathematical models on movement of people in cities. These results were reported mainly in [2]. Other deliverables in the RACE-program also took some of these issues into consideration. In [8] the same basic assumptions were used. The network architecture used was, however, an architecture based on an evolved second generation mobile telecommunication network. In [7], later results from the RACE-program were used and a message service was also included in the study, and these results were used to derive results on loads on nodes and links.

The results in this paper were derived from an analysis of research in a third generation system. These figures are, however, also applicable in evolved second generation systems. The main difference with current systems is the assumption on the traffic model where mobile phones are used in a similar manner to current fixed phones, the assumption that more functions are moved to the telecom databases and also that there are more communication services in the future systems.

The main differences between current second generation systems and third generation systems will be the bandwidth and also that the type of calls will be more diverse. Data communication calls, fax calls and message service calls will be much more common in third generation systems than in current second generation systems where speech connections are the dominating feature.

2   Conceptual Schema of UMTS

In Figure 1 a conceptual schema of the UMTS system is described [1]. UMTS here and in this paper refers to the variant that was developed in the RACE-project, it does not refer to the real variant of UMTS which is not yet completed. Of course in a real system, the schema must be supplied with much more details. This schema does, however, mention the basic entities and their relationships.

There is a set of identifiers and numbers that is used in a UMTS network. SPI is a Service Provider Identity, identifying the Service Provider. ICSI is the International Charged Subscriber Identity: this is the identity of the subscriber, the entity that pays the bill and has the business relationship with the Service Provider. IMUI is the International Mobile User Identifier, which identifies a certain UMTS user, the IMUN is the International Mobile User Number, which is the diallable number of the UMTS user. SI is the Service Identity of the registered service. A terminal address is composed of DI, which is a domain identifier (identifies the network operator) and TMTI, which is the Temporary Mobile Terminal Identifier, a temporary identifier of the terminal in this particular network. LAI is the Location Area Identifier, which identifies a location area used at paging of the terminal. Then we have the MRN, which is the Mobile Roaming Number, which is used for routing purposes, to set-up the connection to the exchange where the user currently resides. Service Provider Data consists of the information on the Service Provider that is needed for execution of UMTS Procedures. This consists of the services offered to home users, services offered to visiting users, agreements with other operators and a list of barred subscribers. Subscriber Profile Data consists of the information on the subscription that is relevant to execution of UMTS procedures, this includes a list of the services subscribed to (at the home operator) and maximum number of users and terminals. User Data contains information on a specific user, his service profile and his authentication and encryption keys. Each user can then have a number of registrations, one for each service the user has registered for. The registration is connected to a specific terminal. The terminal data also contains terminal status and some keys for security procedures.

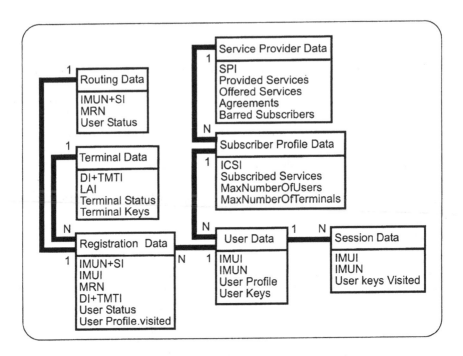

Fig. 1. Data Model of UMTS

It is necessary to have Routing Data connected to the registration to be able to find the Terminal to which the user has registered. When a user is active he also has sessions connected to it. There could be several simultaneous sessions, but there is only one session per service provider, which means that each UMTS database contains no more than one session record per user.

The data in the UMTS network is distributed in databases of the home operator and databases of the visited operator. There are various ways to handle this distribution. A likely way is that the home database stores the Service Provider Data, Subscriber Data and User Data. Routing Data is also needed in the home network to be able to route the call to an exchange in the network the user is visiting. The Session Data, Registration Data and Terminal Data are always created in the visited network; parts of the Registration Data will also be stored in the home network. Parts of the User Data can be transferred to the visited network to make the execution of the UMTS procedures more efficient.

3   UMTS Procedures

The procedures used in an UMTS network will be presented. For each procedure, the impact on the telecom database will be analysed and the queries generated will be shown. Many of these procedures are general and necessary also in another mobile network and most of them also in a fixed network. The information flows of the procedures are based on [3]; in this document these information flows have been mapped to a physical architecture. The major differences of these information flows compared to information flows of current systems is the use of sessions in all communications between the system and the user, the separation of terminal and the user (as in UPT, Universal Personal Telecommunications) and many more services. Finally the telecom database is more involved in the call set-up process where it is involved in both the originating and the terminating part of the call set-up. The network architecture used in the information flows is shown in Figure 2. The SCP (Service Control Point) and SDP (Service Data Point) are normally collocated; they are chosen as separate entities in this presentation, since databases are the focus of this thesis. The interface to the SDP could therefore be an internal interface in the SCP or a network interface. In this way we can derive the number of requests that the SCP database must handle internally.

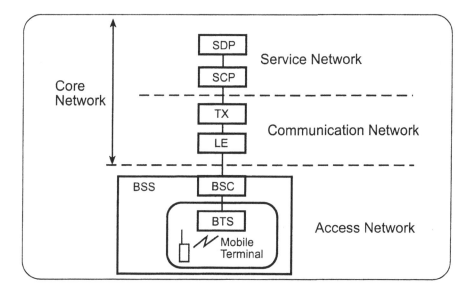

Fig. 2. Network Architecture

All the procedures have more than one possible information flow. For example, a user registration has a different information flow when invoked from a visited network, compared to when it is invoked from the home network. Taking this into account in the calculation of transaction flows in the network would necessitate a detailed mobility model and also that all possible information flows are studied. Such a mobility model was used in [7] and [8]. In this thesis we only show one version of each information flow. This keeps the presentation short and avoids rising an unsure mobility model. The information flows chosen should be representative of the most common cases. All retrieve, update, create and delete operations in the information flows use a unique primary key to access the data, if not otherwise mentioned. This means that retrievals can appear thus:

```
SELECT ATTRIBUTES \\
FROM TABLE \\
WHERE PRIMARY\_KEY = primary\_key;
```

Update, create and delete are performed in a similar way. In some procedures several queries can be part of the same transaction.

In the procedures described below we have focused on the database interaction and hidden most of the other messages occurring in those procedures. [10] contains more details and [3] contains very detailed message flows.

3.1   Session Management

Sessions are used to ensure a secure communication. Before a user is allowed to use any network resource, a session must be set-up. The major reason for sessions is to authenticate the user and the terminal. This authentication ensures that the user is allowed to use UMTS services and also makes it more difficult to use stolen UMTS equipment. Another part of the session set-up is to exchange encryption keys, which are used in the air interface to make sure no one can interfere or listen to messages and conversations.

Since terminals in UMTS do not have identifiers, there are also terminal sessions. The terminal receives a temporary identifier during a terminal session. Services that do not relate to a user can be performed (e.g. attach) during a terminal session. If a user is involved in the service, the terminal session must be upgraded to a user session. Sessions can be initiated either by the network (e.g. at a terminating call) or by the user (e.g. at an originating call). We assume a terminal session exists for a longer period than a user session, therefore we assume that the normal case is that there exists a terminal session during execution of the UMTS procedures.

A session can be used for several communications. This means that sessions are established before calls can be made. However a session can be established for a longer time than the duration of a call. Therefore, if a user makes several calls in a row, these can all be part of the same session. User sessions are normally released a specified time after service completion.

System-Originated User Session Set-up. In Figure 3 the information flow of a system-originated user session set-up is shown. First the database is checked to see if there are any sessions already available (SESS_PAR). Then the database is used to find the user number, given the user identifier (CONV_UI). Then the terminal and the user is authenticated (TERM_AUTH): for terminal authentication an encryption key is needed (RETR_ENC_PAR) from a security database and for user authentication an encryption key and authentication key are needed from the security database too (SEC_M3). Finally the session data is stored in the database (SESS_STORE).

If there already exists a terminal session, then it is not necessary to authenticate the terminal and it is possible to proceed with user authentication immediately. As can be seen, there are five requests to the database, two of which are directed for the security database. Four of these requests are retrievals and one of them is a create request. The retrieval of encryption data for terminal authentication is removed if a terminal session already existed. Also the retrieval of session parameters and retrieval of the UMTS number can be combined into one retrieval request to the database. Therefore there are two retrievals and one create request in this procedure in the common case.

User-Originated User Session Set-up. The major difference compared to a system-originated session set-up, apart from who initiated the session, is that it is not necessary to retrieve any encryption key and it is not necessary to convert the identifier to a number. Therefore there are two retrievals and one create in this procedure. If the user invokes a service when he has a terminal session which needs a user session, then an upgrading of the terminal session to a user session is performed. The information flow for this procedure is very similar to a user-originated user session, the requests to the database are the same.

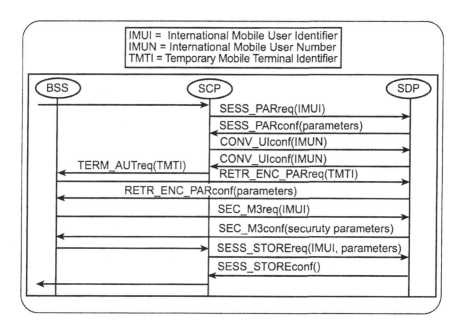

Fig. 3. System-Originated User Session Set-up

System-Originated Terminal Session Set-up. In Figure 4 an information flow of a system-originated terminal session set-up is shown. In this procedure the terminal is authenticated (RETR_AUTH_PAR) arid a new encryption key is stored in the security database (STORE_ENC_PAR) and a new TMTI is assigned

(TMTI_UPD). This inessage also creates a new terminal record. The assignment could be performed with a retrieval and an update in a transaction. This comprises two retrievals, two updates and one create request to the database.

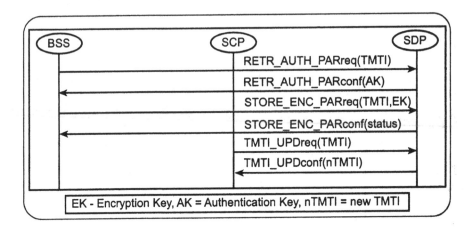

Fig. 4. System-Originated Terminal Session Set-up

User-Originated Terminal Session Set-up. In Figure 5 a user-originated terminal session set-up is shown: the terminal is authenticated using an encryption key retrieved from the security database (RETR_ENC_PAR) and finally the new TMTI is assigned by the database (TMTI_UPD). This message also creates a new terminal record. Thereby there is one retrieval, one create and one assignment in this message. Thus there are two retrievals, one create and one update to the database in this procedure.

Session Release. Finally sessions are released where the session object is deleted from the database. A system-originated session release is performed the same way, only the initiator is different. There is one delete operation in this procedure to perform in the database. When releasing a terminal session the terminal number is also deassigned, which involves one update.

3.2   User Registration Management

User Registration is performed when a user wants to assign a specific communication service (e.g. voice, fax) to a specific terminal. There could be several users registered on the same terminal. Registration is only necessary to be able to receive calls at the terminal. Outgoing calls are performed as part of a session and registration is therefore not needed for outgoing calls. Registration is performed in a session. There could be more than one registration of a user for different services.

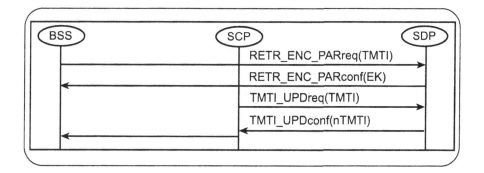

Fig. 5. User-Originated Terminal Session Set-up

The registration procedure (see Figure 6) is one of the more complex procedures in a mobile network. First it is necessary to fetch parts of the service profile of the user to make sure that he is allowed to register for this service (PROF DATA). If the user is registering in a visited network, then the profile must be fetched from the home network, so there is one retrieval in the visited network and a retrieval in the home network to fetch the requested data. Secondly the new registration must be created. This could involve three steps, creation of the new registration (CREATE_REG), deletion of the old registration (DEL_REG) and creating the routing information in the home network (UPD_ROUT). This procedure has some consistency requirements; these have been studied in [4]. They are not included in the procedure shown here.

In the traffic model we will assume that each registration is new and no deletion of the old registration is necessary. In the information flow we do, however, show the full scenario where the old registration is also deleted. Thus there are two retrievals and two creates in a user registration. The user deregistration is simpler, it deletes the registration (DEL_REG) and deletes the routing information in the home network (DEL_ROUT). In Figure 6 we have used SDP-SDP messages between domains. These messages will most likely pass through the SCP-SCP interface in the UMTS standard.

3.3   Attach/Detach

Attach is used to attach a terminal to the network and detach is used to detach the terminal from the network (e.g. at power-off). When this procedure is executed an update request is sent to the database changing the attach/detach attribute of the terminal profile.

3.4   Call Handling

Call Handling is used to set-up, release and modify calls. Call modification is similar to call set-up in what database accesses are needed and therefore we

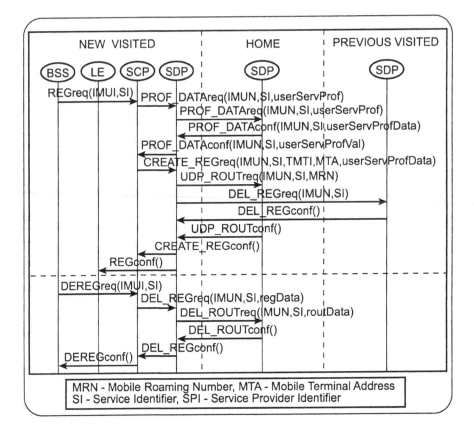

Fig. 6. User-Originated Terminal Session Set-up

only discuss call set-up and call release. Call release is easy, as this does not involve the database at all. In a Call set-up there are two retrievals in both the originating side and the terminating side, which means four retrievals for one call set-up. The first retrieval is performed to check the user profile if the requested service can be used at this time. The second access to the database retrieves the current location of the user. The originating network needs this information to set-up the call to the right network and the terminating network needs more detailed location information to page the user.

3.5 Handover

Handover is performed to change from one radio base station to another; the new base station could be located under a new exchange. It could even be located in a new network. A handover could also be performed internally in a base station between channels in the base station. The database is normally not involved in the handover procedures. Only when a handover is transferring the call to another network is it necessary to access to the security database twice; One to retrieve keys and one to update the keys.

3.6   Location Update

Location Updates are performed as the terminal is moving in the network or moves to a new network. By tracking the users movement with location updates, the user can be called although he is mobile. A Location Update that changes network is called Domain Update in UMTS. This also involves moving the registrations from the old network to the new network. A Location Update updates the Location Area information in the database.

A Domain Update means that a user is moving from one network operators domain to another. As part of this move, all his registrations in the old network must be moved to the new network. This procedure is similar to a user registration. The registration is, however, retrieved from the old domain and it is necessary to delete the registration in the old domain. Some authentication procedure and assignment of a temporary terminal identifier must also be performed. This is only necessary, however, for the first user on a terminal that performs a Domain Update. We assume this is performed by a terminal session set-up before performing the Domain Update. Finally the location data is updated to reflect the new location. If we assume that there is only one registration of the user on the terminal, there will be one retrieval, two updates, one create and one delete during the Domain Update.

3.7   Service Profile Management

The user can interrogate and update his service profile with user procedures. Before he gets this access to interrogate and update the database, the network checks that this request is allowed according to the service profile of the user. After this the user can interrogate and update the database according to what is possible according to his service profile. This will then involve one retrieval and a number of user supplied retrievals and updates. Normally there will only be a few accesses by the user when he manages his service profile. Here one access is assumed per service profile management action.

4   UMTS Characteristics

The most important issue in determining the characteristics of telecom databases used in mobile telecommunication systems is the traffic model. To find the requirements on a specific telecom database, one must also estimate the number of users per database. Current mobile systems have a traffic model which is affected by the high price of mobile communications. In third generation mobile systems the price of communicating with a mobile will most likely be similar to the price of communicating with a fixed terminal today. Therefore calling rates can be estimated by looking at the calling rate in current wireline networks. The calling rate in these networks are 2.8 calls per hour per user (includes both originating and terminating calls) [6]. Since the terminals will be mobile and since so many more services are offered through the telecom network, it is estimated that the calling rate will even increase. There will also be call modifications during multimedia calls that will increase the rate of call handling even more. It is estimated to be 4.0 calls and call modifications per hour. We split this into 2.0 originating calls per hour and 2.0 terminating calls per hour per user. This figure is valid during the busy hours of the day.

User session handling must be performed before each UMTS procedure. There could however be instances where one user session is used for several calls and other interactions with the network. Since user sessions are not used in any current network, this figure must be estimated. We estimate it to be 3.0 per hour per user. It is also necessary to have terminal sessions. These sessions can be active for a longer time, and we estimate that there are 1.0 sessions per user per hour.

Table 1. UMTS Procedure Rates

Procedure	Events/user/hour	Total Events/second	Percentage of Total rate
Originating Call Set-Up	2.0	1111	12.0
Terminating Call Set-Up	2.0	1111	12.0
Inter-Domain Handover	0.05	28	0.0
Location Update	1.6	889	9.6
Domain Update	0.4	222	2.4
User Registration	0.5	278	3.0
User De-Registration	0.5	278	3.0
Attach	0.3	167	1.8
Detach	0.3	167	1.8
System-Originated User Session Set-Up	1.5	833	9.0
System-Originated Terminal Session Set-Up	0.5	278	3.0
User-Originated User Session Set-Up	1.5	833	9.0
User-Originated Terminal Session Set-Up	0.5	278	3.0
Session Release	4.0	2222	24.0
Interrogate Services	0.2	111	1.2
Register Service	0.2	111	1.2
Erase Service	0.2	111	1.2
Activate Service	0.2	111	1.2
Deactivate Service	0.2	111	1.2
Total	16.65	9250	100

User registration is also a new procedure and there are no measurements that can be used in estimating the registration rate. It seems probable, however, that users will register a few times a day. Therefore we estimate that there are 0.5 registrations per user during the busy hour and the same number of deregistrations. Attach and detach we estimate to be 0.3 per user during the busy hour.

The number of location updates is dependent on the mobility model. This model depends on many parameters, such as size of location areas, mobility of users and so on. Also the busy hour for location updates does not coincide with the busy hour for calls. Therefore the busy hour figure should not be used, but rather a smaller figure should be used. The rate varies between 0.3 and 2.5 depending on the conditions. A figure of 2.26 and 2.63 per user per hour is found in a calculation in [5]. We estimate 2.0 per hour per user and of those we estimate that 20% are changing domain (i.e. changing to another operator's network). This gives 1.6 location updates per hour per user and 0.4 domain updates per hour per user.

Table 2. UMTS Database Access

Procedure	Events/user/hour	Total Events/second	Percentage of Total rate
Retrieval	18.65	10361	53.6
Update	4.95	2750	14.2
Create	5.6	3111	16.1
Delete	5.6	3111	16.1
Total	34.8	19333	100 height

Inter-domain handover occurs when the user is moving to another domain of the network while he is performing a call. It is most likely to be very uncommon. To calculate this figure we multiply the rate of moving between domains (0.4 from above) with the probability of making a call while moving to another domain. We estimate this probability to 12%. A very busy telephone line is occupied roughly 12% of the time. To simplify matters we suppose that calling and moving is independent of eachother. This gives 0.05 inter-domain handovers.

The users will also perform service management. It is estimated to rise compared to today, due to many more services. The rate is estimated at 1.0 per hour per user, equally divided between interrogate, register, erase, activate and deactivate service.

The number of users per telecom database depends on how many users there are in the network and the policy of the operator. These databases are essential for operation of the network and therefore most operators will require that the system continues to function even if telecom databases are down. This requires network redundancy, that is, another telecom database is always ready to take over if one goes down. So the minimum number of telecom databases in a network is two. This means that theoretically an operator could have 10-20 million users in one telecom database. However even with redundancy, there could be errors that cause the system to fail. Therefore it seems that 1-2 million users is a more appropriate figure. In this way the large operators can also build their systems with more intelligent backup strategies, where several nodes assist in taking over, when a telecom database fails. In our estimates we are therefore using 2 million users per telecom database. From this discussion we reach the figures shown in Table 1.

Now by using the figures in Table 1 and the information flows from the preceding section we derive the number of retrievals, updates, creates and deletes in the database (see Table 2). From these figures it can be seen that modify queries will put a substantial load on the UMTS database, since a modify query is much more complex to execute compared to a read query.

5 Conclusion

From the description of the interaction between the SCP (Service Control Point) and the SDP (Service Data Point) it is obvious that there is a rather large amount of interaction between the database and the application for each application message received. The reason is that the database is also used to store session information about the telephone call, and also security information. Thereby the requirements on performance and low delays grow even further compared to the requirements by second generation mobile networks.

The requirement analysis shows that mobile telecom systems require databases that can handle a substantial amount of updating transactions. Also these transactions are commonly small (only a few operations per transaction) and require short response time. This creates problems for most commercial database systems. Telecom databases used in the fixed telecom network is usually more readintensive. Thus many studies have focused on building telecom databases with extreme focus on read transactions (e.g. [13]).

Compared to the traditional benchmarking of OLTP applications (TPC-B, TPC-C) which springs from banking applications and business applications, telecom databases contain much less data (typically a few kilobytes per user). This means that telecom databases fit nicely into the main memory of a computer or a cluster of computers. Thus TPC-C is focused too much on disk performance. We are currently performing a test with a main-memory version of TPC-C. We are also going to test the telecom database benchmarks described in [11] to gain more insights on how these different benchmarking categories differ. At Ericsson Utveckling AB we have developed a prototype of a parallel data server for telecom applications, NDB Cluster [9]. From benchmarks on this database we found that it is capable of handling in the order of 2000 queries per second per server using a cluster of Ultra 2 servers (2x200 MHz Ultra Spare) interconnected with SCI [12]. All data in the cluster was replicated. Thus a cluster of 10 servers with 2 added for reliability will be sufficient to handle the requirements of a large UMTS database even with a very tough traffic model. It is even possible to visualise a database which contains the actual state of the telephone call. This state variable would need to be updated several times per call. This would create new possibilities in creating services for telecommunications.

Actually we are currently planning a field trial where a database is connected to a normal telecom switch as an adjunct. The state of the terminals is kept in the database and the database is used to keep a personal profile for personal numbers. This will create a heavy read and write burden on the database similar to the application described in this paper. The major benefit is that it opens up for many new exciting communication applications built with standard IT solutions.

Abbreviations

AMPS:	American standard for analoo, mobile telecommunication
B-ISDN:	Broadband ISDN, a previous standardisation effort
BSC:	Base Station Controller
BTS:	Base Transceiver Station
CDMA:	Radio transmission algorithm
D-AMPS:	Digital version of AMPS
DECT:	European standard for cordless communication
GSM:	European standard for digital mobile telecommunication
IS-95:	American standard for mobile telecommunication using CDMA
LE:	Local Exchange, a telecom switch containing connected subscribers
NMT:	Nordic standard for analog mobile telecommunication
PCS:	An american system for mobile telecommunications
PDC:	A Japanese standard for mobile telecommunications
PHS:	A Japanese standard for cordless communication
RACE:	A European research program during 1987-1995
SCI:	Scalable Coherent Interface, a standard for communication within a cluster of computers
SCP:	Service Control Point, a node in the telecom network providing service logic
SDP:	Service Data Point, a node in the telecom network providing data
TACS:	A British standard for analog mobile telecommunication
TX:	Transit Exchange, a telecom switch connecting other switches to each other
UMTS:	A European standardisation effort for third generation mobile systems
UPT:	An international standard for personal telecommunication services

UMTS Identifiers

DI:	Domain Identifier
EK:	Encryption Key
ICSI:	International Charged Subscriber Identity
IMUI:	International Mobile User Identifier
IMUN:	International Mobile User Number
LAI:	Location Area Identifer
MRN:	Mobile Roaming Number
MTA:	Mobile Terminal Address
SI:	Service Identity
SPI:	Service Provider Identity
TMTI:	Temporary Mobile Terminal Identifier

References

1. R2066/PTT NL/M171/DS/P/061/bl, Implementation Aspects of the UMTS Database, RACE MONET deliverable, 30 May 1994.
2. R2066/ERA/NE2/DS/P/075/bl, Evaluation of Network Architectures and UMTS Procedures, RACE MONET deliverable, 31 dec 1994.
3. R2066/PTT NL/M17/DS/P/099/b2, Baseline Document on Functional Models, RACE MONET 11 deliverable, 31 dec. 1995.
4. R2066/SEL/UNA3/DS/P/108/bl, Distributed database for UMTS and integration with UPT, RACE MONET 11 deliverable, 31 dec 1995.
5. R2066/SEL/UNA3/DS/P/109/bl, Performance Evaluation of Distributed Processing in UMTS, RACE MONET 11 deliverable, 31 dec 1995.
6. LATA Switching Systems Generic Requirements, Bellcore Technical Reference TRTSY-00517, Issue 3, March 1989, See 17, Table 17.6-13: Traffic Capacity and Environment.
7. Larsson, M.; Network Aspects in UMTS, Master thesis at Ericsson Telecom AB 1994.
8. Ronström, M.; Signal Explosion due to mobility, Nordiskt Teletrafikseminarium 1993, Aug 1993.
9. Ronström, M.; The NDB Cluster, A Parallel Data Server for Telecommunications Applications, Ericsson Review 1997, no. 4.
10. Ronström, M.; Design and Modelling of a Parallel Data Server for Telecom Applications, Ph.D thesis, University of Link Sping, http://www.ida.liu.se /eds-lab/publications.html.
11. Ronström, M.; Database Benchmark for a third generation mobile telecom system, submitted for Telecom Database Workshop, http://www.ida.liu.se /eds-lab/publications.html.
12. The Dolphin SCI Interconnect, White paper, Feb 1996.
13. Friedman, R., Birman, K.; Using Group Communication Technology to Implement a Reliable and Scalable Distributed IN Coprocessor, submitted to TINA conference.

Author Index

erg
k
a
ong

re

Databases
n Telecommunications

nternational Workshop, Co-located with VLDB-99
dinburgh, Scotland, UK, September 6th, 1999
roceedings

Goos, Karlsruhe University, Germany

rtmanis, Cornell University, NY, USA

Leeuwen, Utrecht University, The Netherlands

Editor

onker

search

. 15000, 9700 CD Groningen, The Netherlands

w.jonker@kpn.com

ng-in-Publication Data applied for

sche Bibliothek - CIP-Einheitsaufnahme

es in telecommunications : proceedings / International Workshop
ed with VLDB-99 Edinburgh, Scotland, UK, September 6th, 1999.
onker (ed.). - Berlin ; Heidelberg ; New York ; Barcelona ; Hong Kong ;
; Milan ; Paris ; Singapore ; Tokyo : Springer, 2000
re notes in computer science ; Vol. 1819)
3-540-67667-8

ect Classification (1998): H.2, C.2, K.6, H.3, H.4

02-9743
540-67667-8 Springer-Verlag Berlin Heidelberg New York

a company in the BertelsmannSpringer publishing group.
r-Verlag Berlin Heidelberg 2000
Germany

,: Camera-ready by author, data conversion by PTP-Berlin, Stefan Sossna
acid-free paper   SPIN: 10721056   06/3142   5 4 3 2 1 0

Preface

Developments in network and switching technologies have made telecommunications systems and services far more data intensive. This can be observed in many telecommunications areas, such as network management, service management, and service provisioning. For example, in the area of network management the complexity of modern networks leads to large amounts of data on network topology, configuration, equipment settings, etc. In addition, switches generate large amounts of data on network traffic, faults, etc. In the area of service management it is the registration of customers, customer contacts, service usage (e.g. call detail records (CDRs)) that leads to large databases. For mobile services there is the additional tracking and tracing of mobile equipment. In the area of service provisioning there are the enhanced services like for example UMTS, the next generation of mobile networks, but also the deployment of data intensive services on broadband networks such as video-on-demand, high quality video conferencing, and e-commerce infrastructures.

This results in very large databases growing at high rates especially in new service areas. The integration of network control, network management, and network administration also leads to a situation where database technology gets into the core of the network (e.g. in architectures like TMN, IN, and TINA). The combination of vast amounts of data, real-time constraints, and the need for robust operation presents database technology with a lot of challenges such as distributed databases, database transaction processing, storage and query optimization, etc. Finally, there is the growing interest for telecom operators in IP. Both IP based telecom services as well as the integration of IP and traditional networks (like PSTN) require additional database functionality.

With the above in mind, we organized this workshop to initiate and promote telecom data management as one of the core research areas in database research and to establish a strong connection between the telecom and database research communities. In response to the call for papers we were very happy to receive 24 papers both from universities and the telecommunications industry. Given the high quality of the submissions we decided to have the 12 best papers presented at the workshop. The papers have been grouped in four sections addressing network management, service enabling, CDR handling, and real-time databases.

Finally, I would like to address one issue that was also raised during the panel discussion, the question "What is so special about telecom databases?". My answer to that is that the important issue is the extreme requirements that telecommunication applications put on their databases. The requirements are very high in terms of availability, amounts of data to be processed, and number of systems to interoperate. In addition, the telecommunication business is going through a period of extreme growth, especially in the areas of mobile and Internet services, and thus the scalability requirements are very high. At the same time we see specific needs with respect to new functionality and new architectures of database management systems. This requires combined research efforts from the telecom and database communities on key issues such as functionality, architectures, robustness, and scalability.

March 2000   Willem Jonker

Workshop Organizers
Willem Jonker   KPN Research
Peter Apers   University of Twente
Tore Saeter   ClustRa AS

Program Committee
Michael H. Böhlen   Aalborg University
Munir Cochinwala   Bellcore
Anastasius Gavras   EURESCOM
Svein-Olaf Hvasshovd   ClustRa AS
Phyllis Huster   Bellcore
Hosagrahar Jagadish   University of Illinois
Daniel Lieuwen   Lucent Bell-Labs
Stephen McKearney   British Telecom Research
Wolfgang Mueller   Deutsche Telekom Berkom
Marie-Anne Neimat   TimesTen
Matthias Nicola   Technical University of Aachen,
Salvador Pérez Crespo   Telefónica
Berni Schiefer   IBM
Reinhard Scholl   ETSI
Wouter Senf   Compaq/Tandem
Martin Skold   Ericsson
Jeroen Wijnands   KPN Research

Table of Contents